Dairy cows
& duck races

Also from Earthworld:
Mr Trump goes to Washington (Michael Mayor)
Political satire in 3D graphic novel form
Discovering engineering that changed the world
Author Julian Edgar takes you around the world to see an
astonishing series of engineering marvels
Coming soon ...
The Book of the Leica R-series Cameras (Brian Long)
The definitive history of Leica's SLRs
Mum's not the word – Childless, Childfree (Denise Felkin)
A photographic essay challenging perceptions of women
and motherhood

VELOCE PUBLISHING
THE PUBLISHER OF FINE AUTOMOTIVE BOOKS

EARTHWORLD
EXPANDING HORIZONS

Apps
Download now from
www.digital.veloce.co.uk

eBook
Download now from
www.digital.veloce.co.uk

www.veloce.co.uk
EARTHWORLD is a brand new imprint in the house of
Veloce Publishing, showcasing innovative and thought-
provoking books that are informative and entertaining.
Produced to the same high quality of content and
presentation as our existing books, EARTHWORLD
is set to push boundaries and expand horizons.

First published in November, 2018 by Veloce Publishing Limited, Veloce House, Parkway Farm Busi-
ness Park, Middle Farm Way, Poundbury, Dorchester DT1 3AR, England. Tel +44 (0)1305 260068 /
Fax 01305 250479 / e-mail info@veloce.co.uk / web www.veloce.co.uk or www.velocebooks.com.
ISBN: 978-1-787113-79-4 UPC: 6-36847-01379-0.

Dairy cows
& duck races

The life and times
of a young farmer

Philip Dixon

EARTHWORLD
EXPANDING HORIZONS

Author's note

Dairy cows and duck races started life as twenty-six letters that I sent, one every week, to my son who was spending some time working abroad. They told him of a very different world, far away from the one he was experiencing and of my early years in dairy farming. They were all upbeat and, I hope amusing.

To turn the letters into a book I had to fill in the gaps between the stories to link them into a coherent manuscript, which meant baring my soul about all my loves, lost and found. The stories are told with my inevitable enthusiastic poetic licence but all are based on my personal experiences. In a few instances It has been necessary to change the names of certain individuals, whilst certain characters are purely fictional and bear no resemblance to their actual counterparts.

So, read on, "O Best Beloved," of the first fifteen years or so of my working life!

Contents

Peepy Farm

"Harry Dixon the Pharmacist, Purveyor of Fine French Perfumes and Horse Wormers"

So read the sign above my father's shop in Low Fell, a suburb of Gateshead. He was the son of a miner and was born In Bedlington, Northumberland. My mother, Christine Pelagee – "me mam" as I called her – was a nurse from Swansea. They met at the tail end of the Second World War, and, after they were married, my Dad bought a bankrupt pharmacist's shop on Beaconsfield Road, just off the Durham Road. He worked, for what seemed like twenty-four hours a day to make it succeed. If someone was in need, he would turn out, rain, hail or shine, sometimes in the wee small hours, occasionally carrying oxygen cylinders on his shoulders for several miles, because in those early days he couldn't afford a car.

By the time I was ten he had opened several shops and an industrial pharmacy, which sold medical supplies to factory first aid rooms and even once kitted out a Russian hospital ship with a full, up-to-the-minute western operating theatre. We lived in a large house about five hundred yards away from the original shop. He was a great pharmacist, an intuitive businessman, and an enthusiastic philanthropist. He was a good father, but a rubbish dad, because he was never at home. Well, to be honest, that can't be entirely true, because I have three siblings. My brother, Noel, is the oldest. My sister, Jacqueline, is two years younger than Noel, and I'm next; two years younger. Last but not least, the baby, Katharine, (or "Stinker" as I used to call her) almost eight years younger than me.

When mam and dad made me, something in the gene pool must have been playing silly buggers because I have always been the black sheep. My siblings all have scientific degrees and traditional middle class lifestyles, but I was hopeless at school, and, as far as chemistry was concerned, I was a dunce. It has always puzzled me where my love of farming and practicality came from. Certainly my father did not possess any practical ability; he was so bad at anything mechanical that my mother hid any tools he bought because in his hands they became weapons of mass destruction. It wasn't until I was talking recently with an elderly relative that I discovered the inevitability of my career. When I was born, mam told my father to register me as Philippe Charles, but my father, being a northern male, thought this sounded a bit poncy, so it was Philip Charles. I asked why "Philippe," and Uncle Gordon, me mam's younger brother, told me my great-great-grandfather was a Philippe Charles. He was a dairy farmer on Jersey in the Channel Islands. Who says some things are not preordained?

From the age of ten I worked all available weekends and holidays on the Cooksons' farm, west of Kibblesworth, on the outskirts of Gateshead. Working there involved cycling for three-quarters of an hour up and down the steepest hill in the area, for no pay at weekends and bugger all during the holidays. Dear me, I was keen!

I had a place at Auchencrew Agricultural College to study for a Higher National Diploma in agriculture, but as the son of a pharmacist rather than a farmer I lacked agricultural provenance, so before I could take up the place I had to work in a full-time role on a farm for two years to gain the necessary practical experience. So with the help of Auchencrew and before I had left school, a post of what turned out to be slave labour was secured for me on Peepy Farm in Stocksfield, Northumberland. It was a large grassland dairy farm, and the home of the internationally renowned Hunday Herd of British Friesians. For the late 1960s, it was an enormous state-of-the-art enterprise, with automated feeding systems, three times a day milking, and 400 cows. So, at the tender age of fifteen, I left home to start life as a farmer.

I was one of five similar slaves who performed all the menial tasks around this prestigious enterprise. Mr John Moffitt, the second generation owner of Peepy Farm, sorted out accommodation for us by providing one of the larger farm cottages, rent free, to a Mrs Hilda Hodges, to operate a boarding house for the young boys who worked on the farm. My wages were paid directly to Hilda, who, as landlady to us boys, ruled us with a rod of iron, and in exchange for our meagre salaries, did our washing and provided us with all our meals. Rations may be a better word. Once a fortnight, she gave me sufficient pocket money to catch the train home to visit my parents and take Carol, my sweetheart, to the pictures. Carol was my first love and we had been going out together for over a year. For the first year and a half, these trips back home were my social life, apart from the theoretical ten days holiday a year, which we never got, because there was always some agricultural event or something that cropped up, such as harvest, spring work, ploughing, preparing for a show, assisting foreign visitors who came to buy bulls – the list went on and on. It was never convenient for us boys, but we did get overtime and those periods enabled me to start saving.

Hilda's house was already full, so I slept in a caravan under the cover of a barn. The roof of the caravan was the venue for the nightly cèilidh of the Hamish McTavish Rat Clan. In amongst the throng, there was one wise old creature that had no tail and only three legs, the loss of the fourth being a war wound delivered from a previous occupant of the caravan, who had been trying to quieten the rats' noise with the edge of a spade. During the early evening, when the sun was sinking into the west, the slanting rays of sunshine would light up the straw stack next to the caravan and there the rats would soak up the sun, only scattering when they saw me coming. Always trailing behind the disappearing pack was my three-legged friend. When he and a few of his mates clambered onto my roof late at night while I was trying to sleep, it sounded as if there were hundreds of them up there. One night I sat up till two in the morning armed with an air rifle, waiting for them. Not one showed a whisker, but the second I gave up and took to my bed, the little darlings started their merry dance.

My fellow slaves lived in relative luxury in Hilda's boarding house; there were four of them, all very different. Paddy was twenty-two years old with a great shock of black curly hair, big dark brown eyes and a deep olive complexion. He looked like a gypsy and dressed that way too, with a bright red kerchief tied round his neck and baggy trousers, which he prevented from tripping him up by tying baler-band nicky-tams (that's the local name for the straps tied just below

11

the knees of the trousers to stop rats and mice running up the insides of the legs). Then there was Johnny, who was Italian. He could speak little English, but he was as strong as an ox. His party trick was to pick up and carry a full forty-five gallon drum, weighing 200kg, on his own. He would push the barrel over and kneel down in front of it, then roll it up onto his thighs, lean back, inch one foot at a time forward until the barrel was balanced on top of his straightened lower legs and then would waddle around with it. Johnny was the one character that the third of Hilda's residents, Bully-boy Bob, didn't mess with. Bob was about nineteen, lean, mean and well muscled and had the early stages of small man syndrome. Finally there was Tommo, who was just an ordinary sort of chap doing his two years practical experience before going to agricultural college.

We never gelled as a group or became great friends because we all worked different shifts, were mentored by different people, and rarely worked together as a team. As a group we were the lowest of the low of the farm staff and consequently we had underdog camaraderie rather than friendship. Having said that, Paddy, with his Romany background, taught me to tickle trout, and one evening after work we successfully pulled six good-sized trout out of the stream that ran past the steading (that's the northern agricultural term for the main group of working farm buildings). Paddy taught me another valuable trick. One day when we were cutting dyke backs (hedgerows) by hand to make the farm look pristine for our international visitors, I cut a wasps' nest in two. I was instantly surrounded by a swarm of very angry soldier wasps. The normal reaction to this was to warn everybody else by shouting "WASPS!" and legging it as fast as possible, but you'd never be quick enough and would always get stung several times. Paddy showed me how to spot and kill the queen. As long as you found her quickly, this instantly dissipated the swarm and you got away with very few stings.

On my first day at work I was subject to the sort of tricks that **are** played on all young lads in the same position the world over. First, I was sent to the workshop for a cap full of nail holes. There, the workshop foreman snatched my hat from my head, and to my dismay put it on the wooden bench and nailed it to a piece of wood with five nails. Then

I was sent to the herd manager for a "long stand." He said, 'wait there.' So I did. It took twenty minutes before I cottoned on. Fetching a churn full of steam was my next errand. I was sent to the dairy with an empty churn and there I asked the dairyman for my steam. With great showmanship and panache he filled my churn to overflowing with steam from the sterilising machine. Then, with a big clang, he clamped the metal lid firmly in place, trapping the steam inside. I dutifully took it back to the workshop where the foreman, on opening it, found that the steam had condensed on the cold aluminium of the churn. I was sent backwards and forwards five times before I twigged they were having me on.

On my second day I was taught bull walking. This task involved training young, and sometimes not so young, bulls to walk on a halter. This was an essential job because the Hunday bulls were shown at Agricultural Shows all over Britain and one man had to be able to lead his charge around a public show ring. At that point in time, Britain was the stockyard of the world. The show-stopping Hunday bulls were sold all over the globe, so winning awards like Best in Show would add thousands of pounds in value to the bull and increase the prestige of the herd.

To teach me the task, one of the senior stockmen took me down to the bull pens. He haltered one of the younger bulls, which weighed about 400kg, handed me the loose end, telling me that he was worth £60,000 and on no account was I to let him go. He opened the gate of the pen and the bull exited like a bullet out of a gun, with me being dragged behind, across the open yard and into the muddy paddock opposite. I was vaguely aware of the roar of approval of all the assembled staff. I hung on like a water-skier who's lost his skis, just managing to keep my head above the wave of slurry that had been conveniently placed in the gateway for my added enjoyment. The staff had organised a book on how long I could stay attached to the rope, but much to my relief and the disappointment of the spectators, I never had to let go. I was such an insignificant weight for the bull to tow that he wasn't aware of me and he came to a standstill at the nearest tuft of fresh grass and started to graze.

After an initiation of this sort I was certainly ready to pay attention when my elders taught me the correct way to walk bulls. Harold Hodgeson, the Hunday herd's manager was smaller than me. He had skin like well-chewed leather but possessed legendary stockmanship skills. On my third morning, he was waiting for me outside the bull pens.

'First of all, young man,' he said. 'Show no fear. A cool head and a calm confidence is half the battle.'

We then went into the bull pens area and I got a good look at twenty-odd bulls held in pens that were built like German gun emplacements. I thought, 'Oh, shit!' These creatures were all bloody enormous.

Now there are several tricks of the trade when walking a bull. The first is to get in very close; in fact, you should be touching the animal, because a hard push from a bull is always less painful than a full-on head butt. The next is that by law, all dairy bulls have to have a stout ring through the soft tender central area of their nose. When you walk a bull, your right hand holds the halter just behind his jaw and you jam your shoulder up tight in front of his horns. You then feed the six-foot length of lead rope through the nose ring, then, using your left hand you grip each side of the rope so as to form a short handle with the nose ring clamped in the middle. This allows you to twist the rope, which in turn twists the nose ring. As I am sure you might imagine, bulls do not like having their noses twisted. It hurts.

These detailed instructions are all well and good, but first, you have to catch the monster. You do this by slipping the halter on while he is feeding, then position yourself as quickly as possible hard up against him with his nose ring clamped tight. Before the pen gate is opened, you walk him round and round the confined space of the pen, twisting that nose ring hard if he starts to misbehave. At this point someone on the outside of the bull pen will deem you are in control and will open the gate. The bull thinks 'freedom!' and 500kg of pure beef tries to head for the hills. I weigh 80kg now, but back then I weighed no more than 60kg, so you'd think it would be no contest, like it was on my second day, but with confidence and experience, believe or not, the technique works. The bull may dance around you a bit but on the whole you stay in control.

As the months pass and confidence builds slowly. So does the size of the bull you are entrusted with. All young testosterone-driven boys quickly become over confident, even with a monster weighing nearly 1000kg, and I was no exception. Walking one of these beasts gave you a rush like nothing else, but as months progressed and the adrenaline no longer rushed like it did for your first orgasm, it was time for your baptism of fire.

All bulls are potentially dangerous and it is always the quiet ones that have never hurt a soul that get you. Cattle arc divided into two categories; beef and dairy. Beef cows and bulls are very similar looking – stout, beefy and not too hormonal. Dairy cows and bulls are the complete opposite of each other. The cows are extremely feminine, complete with PMT and just about everything else that goes with the gender. The bulls are raging testosterone-filled sex machines. Jersey cows are the most feminine of the bovine genre, with big blue eyes and dark lashes, as if someone had painted black kohl eyeliner to emphasise them. They are slim, dainty and have shapely udders. But in typical female fashion, if they don't want to do something, they will lie down where they are in a complete huff and will only move when they decide to. Jersey bulls are permanently in a bad mood, have pointed horns, are lithe and fit and have a turning circle of minus three feet.

My baptism of fire took place on a farm belonging to Mr Moffitt's cousin, who had a large herd of Jersey cattle. I was sent there to help train the new stud bull for the herd, or so I was told. Like any young man I was keen to show off my skills and when I saw the miniscule size of my pupil, which weighed about 250kg I thought it would be like taking a French Poodle for a walk in the park. How wrong can you be? I was about to get a lesson in "never judge a book by its cover – or its size!"

The bull's name was Marmaduke and he lived in an old stable. He could barely see over the split door, but his long curved horns were clearly visible above it. He was already wearing a leather halter and it only remained for me to attach a lead rope. At this point he was very cooperative and I quickly clipped the rope onto the halter, threaded the open end through his ring and tried to pull him tight up against me. That was a problem, because his horns stuck outwards and upwards, much like the cow with "the crumpled horn that tossed the dog that chased the cat that lived in the house that Jack built." There wasn't enough space for me to stand in front of the horns and I would be far too close to their points if I stood behind. So what did I do? Because he was so little and I was the Big Man, I gave him space. As soon as we left the confines of the box, Marmaduke felt the looseness of the rope and thought he was free. He shot forward until the lead tightened, then he span on the spot to face me. Dropping his head, he took a step forward, put his nose between my legs and tossed me onto his back. Then, as I lay face down, spread-eagled across him with my limbs unceremoniously draped across his rump, he bucked up and down several times. Each buck banged my jaw closed, making me dizzy, and eventually I slithered to the ground. This episode only lasted about thirty seconds, but as my senses cleared I became aware of a cheering crowd who had appeared out of nowhere to witness my humiliation.

I jumped to my feet and approached my quarry for a second time, trying my hardest not to limp, because the little bugger had clipped my thigh with his horns. My chin was none too happy either. Marmaduke did not run away. He just waited for me to try again. Did I see a smile on his face? This time I was more wary and I made a plan to keep ahead of this lithe, bovine Jacky Chan. Unfortunately for me, so had Marmaduke. As soon as I had gathered the rope and was moving in close, he moved closer and dropped his head, but kept it tilted back. Then when he was touching me, snapped it forwards, dropped his nose, trapping me between his horns. He stepped forwards, flicked his head and over I went onto his back, but this time backwards. I rolled down the centre of his back and landed on my feet at his rear end, like some Spanish *banderillero*. To my shame, the crowd cheered again. As I had passed over his head, the lead had got wrapped around one horn. This left only a short piece dangling for me to get a hold of, and he knew it. Each time I approached he stuck his nose in the air and every time I was close enough to catch hold he would move it, just out of reach. He was enjoying himself and the crowd thought it was hilarious.

After I had missed it about seven times in a row I had the bright idea of grabbing his two horns, much like you would the handlebars of a bike. With the extra leverage the horn length gave me I managed to twist his head sideways. Marmaduke clearly thought this was cheating, because all of a sudden he turned on me, bellowed and hit me midriff, sending me over his back again. But this time he poked my hip with his horn as I flew past. As soon as I landed he turned to have another go, but luckily the onlookers intervened and shepherded him back to his stall. I limped for a few days afterwards and breathing was painful for a fortnight. Even now, when I am ill the scar on my hip that Marmaduke's horn gave me appears mysteriously, like some ghostly apparition.

There was a definite pecking order in amongst the boys and being the newest, youngest and the weakest, I was a prime target for bullying. This mostly manifested itself as being given the worst jobs, rather than actual physical abuse. However, one sunny August day while we were forking bales of straw and heat exhaustion was fraying our tempers, Bully-boy Bob ordered me to fetch water for him. This was once too often, and I told him to get his own water.

Silence fell on our group of five: I had dared to challenge the self-imposed leader of the slave gang. Three of us had been stood around the long yellow bale trailer with long bale forks in our hands. The other two were on the bale trailer stacking the first course. Bob's mean face twisted into a snarl, he stuck the hay fork into the ground, tore his sweat-stained shirt off his well-muscled torso and stood glowering at me like a miniature Arnold Schwarzenegger. He paused for about ten seconds while his temper got up to a full head of steam, then he put his head down and charged at me. I was so tired I did not react until the second he had launched himself into the air in a flying head tackle. I would like to be able to say that what happened next was a move I had learned on the streets, but the reality was that I turned and ducked as Bob flew straight over the top of me. He didn't quite make it and his feet caught my shoulders as I started to straighten back up. Now upside down, the small of his back smashed into the sharp metal edge of the trailer and he slumped down, bumping his head on a large stone that was sticking out of the ground.

When his body came to rest I thought I had killed him, but when his screams started, I thought I had broken his back. I had done neither, but he had a lump the size of an egg growing out of his forehead and a bleeding bruise across the small of his back. I just turned and stretched out a hand to help him to his feet. Groaning, he struggled to straighten his back and clamber upright.

'Ye di'ne tell us 'bout the karate,' he grumbled. I just shrugged and got on with the bale lifting, while he went to the corner of the field to nurse his wounds. We were never friends, but from then on he was always wary of me and kept his sneering well out my earshot. The non-existent karate training was never mentioned again.

Life at Peepy Farm was physical graft from dawn till dusk, be it helping to milk the large dairy herd or cutting dyke backs with a sickle, but Saturday mornings were always set aside to sweep the yards. These were like motorways running around the circumference of the steading and it took our slave gang a good four hours to perform the task. One Saturday we were half way down one side of the building sweeping in line, each of us cleaning a strip about seven feet wide, when John Moffitt joined us to see how we were getting on. He insisted the brush heads were flat to the floor, using long brush strokes, and assured us that we didn't need to stop to talk because he could walk and talk at the same time and he was sure we could do the same.

It was at this point, at the far end of the yard, Maky appeared.

Maky was a fully-grown Dutch Friesian bull, with horns, a large hump and a temper like, well, a raging bull. He was up on his toes, trotting and moving sideways as he rounded the corner and into view. He had saliva dripping from his mouth, snot trickling from his nostrils and stuck across his horns was a twelve foot long metal gate. He had obviously broken free from his tether and had made his bid for freedom by charging at the gate and managing to create this rather large tiara.

So, if you would picture the scene: there were us five young boys and John Moffitt, all in a line with yard brushes in our hands, and all about a hundred yards away from Maky. To our right there was the solid wall of the cattle sheds and to our left, a large opening into a covered straw yard. In a strong, assured voice John Moffit said, 'hold your ground lads, and we'll turn him into the straw yard.'

Maky had come to a standstill as he assessed the situation, he pawed the ground and tossed his head up and down which made the gate bang up and down on his head. He bellowed in fury and set off towards us, first at a steady trot, then a canter, then, in the blink of an eye, into a gallop.

With each thunderous gallump, the gate banged down painfully on his head.

'Hold firm lads, he will turn.' John said, but with not quite as much confidence as before. We were not about to move. We were frozen to the spot, mesmerised by terror.

'I'm sure he'll turn,' said John.

With the advancing mountain of meat, horn and gate, the five foot high solid stone wall about sixty yards behind was becoming more and more inviting. One by one, we boys gave up and took flight towards the wall. There was already a figure ahead of us, yelling over his shoulder, 'Come on lads, quickly, over the wall!' It was John Moffitt! He had chickened out before all of us boys!

Maky hit the wall with a clatter and a thump, but it held. Then he turned and wandered away, snorting and dripping foamy sweat onto the ground. This was when Harold Hodgeson, cow whisperer and Hunday herd superman, appeared. He walked straight up to Maky, and with a few tugs, lifted the gate off his head. He then latched two fingers into Maky's nose ring, turned his back on him, and led him to a secure bullpen as if he was leading a naughty child. Was this confidence? Was it foolhardiness? Or was he a man with a true stockman's eye? Like horse whisperers, they are a rare breed, and although the father and son team of Moffitts contributed the acumen and received the accolade for creating the Hunday Herd, it may be said (and I could not possibly comment) that Harold was the true talent.

All of us boys had the hots for John Moffitt's seventeen-year-old daughter, Susan. The competition between us to gain her attention when she walked around the farm buildings made us act more like giggling teenage girls than young, fit, bronzed, well-muscled (or so we thought) men, but the sight of her in her skin-tight jodhpurs and knee-length shiny boots was too much for us group of baboons. We would wolf-whistle, hoot, bark like dogs – anything to gain her attention, but in response to it all, she would tilt back her head and put her nose in the air. We all thought that she was being a snob: we were staff and therefore weren't fit to lick her boots. Judging by the way we behaved, however, I think she was justified.

A couple of times I did have the opportunity to talk to her on a one-to-one basis, but just as I thought it was going well she would revert to being a teenage girl and cut you to the quick by saying something really bitchy. Was she a snob or was she spoilt? Looking back, I see her behaviour as that of a normal teenage girl, but it was the start of one of my life's big learning curves, that of trying to understand the complexities of the female of the species.

As I was approaching the eighteen-month mark at Peepy I realised that once you were employed as a boy you were always a boy. Paddy had been there since he was fifteen and had not progressed and I had stopped learning anything new. I was also aware what an enormous gap there was in my practical agricultural knowledge. I had just turned seventeen and I was miles behind farmers' sons who had started learning their trade as soon as they could walk. I needed a full hands-on job, so I started reading the ads in week-old copies of the *Farmers Guardian* that were left in the workshop when the big house had finished with them. I noticed an ad from a farmer near Morpeth, who required a tractor man/relief milker. I applied.

North Farm
and Tom Hall

T R S P Norton – Timmy to his friends – was joint master of Morpeth Hunt and a senior officer in the Northumberland Fusiliers Territorial Army. He very much played the part of a country squire and banking type – 'live in Whalton Manor, don't you know?' Although he seemed ancient to me, he was probably only in his late thirties and he looked the part, with his slicked back hair and his crisp, stripy shirt and matching tie, and he drove an Alfa Romeo GTV. He was newly married to the young, good-looking, blonde Penny. They owned North Farm and the Home Dairy; nearly 400 acres in total and a herd of one hundred Ayrshire cows. There were 150 acres of grass and the rest of the estate was given over to arable crops, mainly wheat, barley and oats. Although Mr Norton owned the farm, he had delegated the running of it to the local land agents, Campbells Green. Their representative was the reprehensible Mr Leon, who had in turn employed Mr Tom Hall as the farm manager.

My move to North Farm happened incredibly quickly. I was interviewed by Mr Hall in his kitchen one Saturday morning and within ten minutes he offered me the job. He told me I would be paid at the Agricultural Wages Board Rate (farm workers wages were government controlled back then) and there was a two-bedroomed cottage that went with the job. I was so inexperienced and nervous I didn't ask any questions or ask to see the house that was to be provided.

He told me to hand my notice in to Mr Moffitt on the following Monday morning and move into my house the following weekend. I was paid weekly at Peepy so one week's notice was all that was required. During the following week I walked down to the local garage in Stocksfield, handed over most of my savings and in return became the proud owner of a 1956 low-mileage Austin A55. I would have been four years old when it was first registered. Inside it smelt of leather and oily engines and it turned out to be a superb workhorse. When the Saturday arrived, I drove it to Whalton together with all my worldly possessions.

Mr Hall met me in the centre of the village and showed me my new house. It was attached to Timmy and Penny's very beautiful Manor house, which was a T-shaped building, with the front, which was the main part of the house, running parallel to the main road. Behind this, the old servants' quarters formed the upright of the "T." Starting nearest the main house were the manor's kitchens and storerooms, followed by the housekeeper's quarters and then my tied cottage, which was embedded into the raised formal gardens. It was a stunning location. However, because my caravan at Peepy Farm was furnished, I didn't have a bed, a chair or any pots and pans to cook in. I did have a suitcase full of dirty clothes (when Hilda heard I was leaving, she stopped doing my washing) and my well-scrubbed wellies. The newly renovated cottage had a brand new fridge, washing machine and a cooker.

Having deposited my case in the sitting room I drove into Morpeth, and at the camping shop I kitted myself out with plastic crockery, a set of pots and pans, a camp bed, a sleeping bag, two folding chairs and a camping table. Returning from the metropolis I set up my meagre belongings and realised I had forgotten to buy something to put into my fridge or into the shiny new collection of camping pots and pans. The village shop had closed and so there was nothing for it but try out the scran (that's what us Geordies call food) at the Berrisford Arms. At this point in time I had not told my parents I had changed my job and moved.

My taking the post of slave boy at Peepy had been a joint decision between Auchencrew Agricultural College and my parents. I think this was my first independent life choice and I had taken it without any family consultation. As I sat in the pub, devouring my steak and kidney pie, I thought about the phone call long and hard, and I came to the conclusion that at about 9.30 on Sunday evening would be the best time to call. There would not be time for my parents to change anything before my start time of 7 o'clock the next morning. So, at the given hour, under the streetlights in the red phone box in the centre of Whalton, I dialled the number. With the icy hand of worry gripping my stomach, I pressed button "A" and connected with my parents. As it happens I needn't have been concerned. My father answered the phone, and, after I had told him he said, 'Well done and good luck,' and then said good night.

However the next week, when I rang at a more normal hour, me mam answered! 'How could you?' She said. 'You've given us no address! Where are you living? Supposing something had happened to you?' And all the usual mother stuff.

My first morning started at 6.30 with a brisk walk up to North Farm where Mr Hall (to me he will always be Mr Hall – I still can't bring myself to call him Tom) and his wife, Joan lived. It was a traditional Northumbrian farmhouse set well back from the road and on a hill just outside the village of Whalton. They had two sons, who were by then young men and had flown the nest, leaving behind mum and dad with Whisky-Sue, a blind golden retriever and Sixpence, a feisty Jack Russell.

Mr Hall turned out to be my agricultural father and my first one-to-one boss, and he never commented on my youthful audacity for trying to con him into thinking I was more experienced than I was. However, Mr Hall's wife, Joan, had arranged for my midday meals to be cooked for me, so on reflection, they were well aware of my youthful inexperience. While working for Tom Hall I got a grounding that has served me all my life, from how to approach and solve seemingly impossible situations, to kicking my ankle when driving a tractor so I would not ride the clutch.

'Clear the decks and let the dog see the bone,' he would say when one or both of us was in a fix, like the one where I was standing scratching my head looking at a broken muck spreader buried up to its axles, jammed up against a six foot oak gate post in the middle of a muddy opening.

Mr Hall was not a big man, but he had a stocky presence. I suppose a taller, slimmer version of Harry Secombe would describe him, and he had a voice to match. When we had finished milking and routine cleaning was taking place, his dulcet tones would be rendering, *I'll Take you Home Again, Kathleen.*

Dapper is too strong a word to describe his appearance, but he was always clean, neat and tidy. If M&S wanted a model for a range of countryman's clothes, he would fit the bill. To complete the ensemble he nearly always wore a tie or a kerchief, plus a flat cap while at work to protect an increasingly receding hairline. He would turn his cap back to front during milking, so that he could get close to the Ayrshire cows to put the milking cluster on without it being pushed off his head. When he milked, he rejected the normal milking apron that most dairy farmers used to keep the cows' wet sloppy muck off their overalls. Instead, he chose to wear a vet's calving apron. These had cap sleeves with strong rubber seals that clamped around the biceps and then swept down to the ankles in a cascade of rubberised material. They fastened at the neck and down the back and covered everything. James Bond could have stepped out of one of these after milking a hundred cows, his dinner jacket looking immaculate. OK, that's an exaggeration, but you get the idea of how protective they could be. Or almost.

One spring morning when the grass was lush and nutritious and the herd had gorged themselves on it, we were milking in the herringbone parlour, standing with our chests level with the cows' udders. We were about half way through milking and Mr Hall was struggling with a new heifer (that's a two-year old cow). She was not enjoying her swollen udders being interfered with and she was swishing her tail around in irritation. Normally we would put the units onto the cows from the side, but because this one was small, she had moved forwards and squeezed in behind the cow in front. Mr Hall had no alternative but to put the unit in between her back legs and hold it there while she milked. She was not happy and continued to swish her tail around, knocking his cap off and sending it flying across the parlour, where it landed in a heavily soiled drain. With no hat, his head and the back of his neck were completely exposed and his apron was gaping away from his back. He was bent over in this position when, like a well-directed fire hose, a hot steaming stream of slurry exited from the heifer's backside. It hit his head and his neck, then inside his shirt. By the time she had finished it was running down his trouser legs and into his wellies. He looked like he had been held upside down and dipped in chocolate.

He was so annoyed; he was stamping his feet like some demented Rumpelstiltskin. For me, solving this problem was easy. Well, he'd helped me out of enough tight spots, so now was my chance to return the favour. I stood him in the middle of the collecting yard and power hosed him down with cold water. He stood there, peeling off one washed layer at a time, turning them inside out to make sure everything was cleaned. Eventually he was down to his Y-fronts, blue with cold, with no dry clothes to wear to drive the several miles back to North Farm. I was very sympathetic to his plight, which was easy to imagine, but difficult to see through the tears of laughter coursing down my cheeks. I sat him on an old plastic fertiliser sack in his beloved new 1600 VW Beetle and he drove home, leaving me to finish up on my own.

At North Farm, the steading lay behind the house, and beyond its quadrangle of attractive, traditional stone-built stockyards and stables lay a modern concrete Atcost building. This housed the large corn dryer, grain bins and straw yard, which from September onwards when we had completed that year's harvest were all full to the rafters. We always sold the grain in January and February to get the best prices, but the downside of this was during the four months when the grain was in storage, rats and mice were able to live in the straw and forage on the top of the grain bins and no matter how much poison you used to bait the traps, very few rats were eliminated. They bred and ate, and bred and ate in a rat paradise; unlimited food and warm, dry straw to make their nests.

The winter's day we emptied the silos was very bad news for our rodent friends, because when they made their cheery way to lunch, the ground they were so used to walking on was no longer there and they tumbled forty feet down to the bottom of the grain bins. You would have thought a fall of this height would have killed even the hardiest of rodents, but they actually had a soft landing on the sloped pile made up of the grain that did not make it out of the six-inch outlet when the silos were emptied.

To remove the last of the grain, it was necessary to get into the silo via a manway, which was situated at the base, and shovel the last few tons to the outlet located directly below the manway. There are two dangers with this procedure. If the corn had got damp in any way it would stop being inert and would start germinating, producing carbon dioxide. This is heavier than air and settles in the bottom of the silo and would suffocate you if you got into the bottom of the silo and couldn't get out. The second problem was if you threw open the hatch and climbed in you would be greeted by hundreds of rats in a state of confusion and desperation. So, that was your fate; be drowned in a sea of CO_2 and then be well chewed by desperate and terrified rats. Not good.

Mr Hall solved both problems. First he would crack open the manway, not so far to allow our furry friends to escape but far enough to allow any residual CO_2 to spill out and dissipate overnight. This had the added benefit of allowing the maximum number of rats to tumble into this giant trap, and by morning there were often more than two hundred rodents squealing and squabbling in the bottom. The next part of his solution involved Sixpence, his Jack Russell, who jumped voluntarily into a 25-litre yellow plastic bucket and was lowered into the bin on the end of a rope. She was so up for this that while we were tying the cord to the bucket handle, she kept jumping in and out of the bucket, yapping and dancing about. Sixpence was so keen to get to her work that she would often jump out the bucket before it reached the bottom. It only took her a short while for her to sort the problem out, then we would open the manway and go in armed with empty fertiliser sacks to clear up the warm, grisly aftermath, picking them up by their tails, if Sixpence had left them one. We often filled up two or three sacks. As for Sixpence, she was bloody but unbowed, with well-ragged ears and several bites around her torso. She was doing what she had been bred for and she loved it.

Mr Hall taught me the stuff that wasn't in the text books and the skills that I fear will be lost for all time because of economic pressure put on the industry by the big supermarket chains. He taught me things like a "stockman's eye," which was the skill Harold Hodgeson had put to such good use for the Hunday Herd and was essential if you were going to work with animals. Some of these skills we can be taught, but some of them we are born with, to a greater or lesser degree. The beginning of this part of my education happened one day in the milking parlour when we had a small electrical problem and the metalwork in the pit was mildly live. We just felt a slight tingle, but cows are very sensitive to electricity. Where it takes several amps to kill a human, it only takes a few milliamps to finish off a cow. Their reaction to the live metalwork did not take any skill to notice. They were going berserk.

When Mr Hall found the cause (which was a loose wire touching the metal stall), and the cows appeared to have settled, he told me to go into the collecting yard. 'Smell the air,' he said. 'Then tell me what you can smell in it.'

I thought he was having a laugh, so I told him it was what you'd expect in a confined space holding over a hundred cows. It smelled strongly of cow shit. He sent me back in to have another sniff, and he was right. In amongst the pungent smell of ammonia there was another acrid smell that caught in my throat. That, he told me, was the smell of fear. The cows in the parlour that had been electrocuted gave off this smell and the alarm signal it gave had spread to the rest of the herd. Within the confines of the enclosed yard, that nasal trigger grew progressively stronger and that in turn sparked an elevated adrenaline response in the herd, working them up to fever pitch.

In one way or another over the next seven years while we were milking together or handling the cattle he managed to fine tune my senses to recognise a whole range of signs that would help me look after my future charges. It included body language, behaviour, sounds, breathing and smells, recognising points in breeding cycles, sickness and mood swings. And yes, cows do have emotions. If you don't believe cows have emotions, try standing between a cow and her calf with your dog on a lead.

I now think these smells are pheromones and it just took somebody to point out what each of these signals meant. Over the years with this basic grounding, my abilities have sharpened, and today I have some strange abilities that seem to work around my friends and family as well as the farmyard. They seem to be at their keenest when I am living in the countryside, where my senses are not bombarded by the industrial stench of the urban environment.

Mr Hall was a social character and got out and about regularly, visiting friends for dinner parties or fun weekends. He reciprocated by having a big barbecue and cèilidh once a year. Cèilidhs and barn dances were, and still are, very popular in rural Northumberland. In fact I can't think of any occasion whilst living North of the Tyne when there wasn't country dancing and of course there were lots of reels that Scotland has claimed the origins for, despite having their roots firmly planted in the borders. The preparation for Mr Hall's annual bash was held in the fold yard, a lovely stone and blue slate affair opposite the farm house. This is where we held the sterkies (yearling heifers) in the winter and the dance started soon after we had put the kine (that's an old English term for a herd of cows that's still in use in Northumberland) out to grass at spring time.

To get the place ready, we opened up the covered five-bar gate and removed the four-foot deep FYM (farm yard manure) with a tractor loader and a seven-ton trailer, taking it to a midden, well away from the farm. The following year the manure, by then well composted, would be the main fertiliser for growing corn. Every little bit of debris was scraped from the walls and out of the corners of the holding yard. It was then left open to dry off and after a week or so, we returned, armed with buckets, brushes and a big sack of whitewash. Normally we liberally sloshed on two coats of whitewash to the walls, not bothering to clear up the splashes all over the floor. This left the yard smelling sweet and looking startlingly bright white. The yard was then used as our temporary storage to house a batch of wheat or barley that was a bit wet or a good batch of barley that was going for malting. Of course it was necessary for the yard to be clean for this, but white?

After the corn was gone we would get the Burco boiler working, clean out a black bin and start making the Orkney Brew for the party. This is a fabulous summer beer that slips down like lemonade and kicks like a mule. The recipe, which was given to me some time later, calls for dozens of clean beer bottles to be ready for the final stage, but for the life of me, I can't ever remember us doing this at the farm. All I can remember is dipping glasses into the bin and drinking it. We missed out the bottling stage completely.

Next we prepared the barbecue. Ours was made of an old empty oil drum cut in half top-to-bottom and mounted on angle-iron legs. We scrubbed the inside of the halves with wire brushes to remove the rust and fatty debris from the previous barbecue. When the time came to fire them up, we charged them with oak chips and lit them, and when the smoke died down, filled them up with charcoal and laid a thick steel grille across them. They would stay hot all night with very little attention and one hundred-plus people cooked their supper on them. We wired up the fold yard with enough cable to drive the lights and, more importantly, the amps, mike and lighting for the band that played the reels and polkas for *Strip the Willow, Dashing White Sergeant, The Gay Gordons* and the like, into the wee small hours.

I can't recall it ever raining on any of these parties. They always started in a field beside the house with a game of rounders with everyone joining in. Boy, did you have to hit it hard to score a round when the fifty or more guests were fielding. So it was an evening of drinking hard, eating well, energetic playing and dancing. I actually do not have any memories of ever leaving any of the parties but I must have done because I can recall crawling back to work the next day.

It was shortly after one of these dos that we had a new sprayer delivered. The flow rate through the main nozzle was not big enough, so I was despatched with my tractor to Jack Young, the blacksmith at Cambo, armed with a blank brass nozzle and written instructions, which read; "we need a hole drilling In this sufficient to allow a flow rate of 200 gallons an acre at 4mph through 16 nozzles at 200lbs per square inch."

We had used Jack in the past to straighten metal girders with his forge and a big hammer. He was six feet, four inches tall and a lean, sinewy man. He wore a dirty, blue bib and brace overall that he had never changed and he didn't seem to wear anything under It. One black, rotten tooth stuck out of his mouth (for this, we nicknamed him "Juanita" – one eater), and beneath a filthy flat cap was a dewdrop hanging on the end of his beaky nose and a Woodbine, super-glued to his lower lip. He was a man of few words but many grunts and mutterings. I couldn't imagine this man having the academic qualifications to solve our problem. Instead, I expected him to give me explicit instructions as to where I was to insert my blank brass nozzle.

I was so wrong! From out of his back pocket he took a fag packet, and from behind his ear a sharp pencil. He licked the pencil, gazed heavenward, frowned and started the calculation. Muttering and scribbling down formulae, asking how many additional nozzles it was to serve, how fast the tractor would be travelling, what pressure the tank operated at etc. Then using πr^2, the coefficient of friction, pressure, the cumulative area of the nozzles, the ground speed and other calculations I hadn't even thought about, it took him about ten minutes to arrive at his answer. He used no computers or calculators; just an agile brain and he must have been paying attention at school. But most amazing of all to me was to realise that algebra was actually relevant. I just stood there awestruck, my mouth gaping.

When he had finished he put the pencil back behind his ear and returned the fag packet to his back pocket. He then turned his back to me and walked over to his work bench, selected a drill bit, inserted it into his pedestal drill and drilled a nice neat hole in the centre of the blank nozzle, scattering golden coloured brass swarf onto his bench. When he was happy with the hole he gave it back to me with a 'reet, that'll be it.'

That was it from him; no unnecessary words, and pretty well typical of all blacksmiths I have had dealings with. Without exception they are all as mad as fish, and are all ruled by the moon.

Armed with my prize, I returned to North Farm. Mr Hall and I fitted the nozzle to the sprayer and we charged up the tank with water to test it. It was miles out; the hole was far too small. This, however, has never diminished my admiration for Jack Young of Cambo. He was perhaps forty years my senior and perhaps he got the maths wrong, but he knew how to calculate a complex problem using algebra. Of course I am also aware of the alternative – that he had a cracking sense of humour and I had been suckered.

ORKNEY BREW

Ingredients
6lb of extract of malt
8oz hops
12lb of caster sugar
8 gallons of water
1oz of brewer's yeast

Equipment
1 plastic dustbin with lid
1 Burco boiler to boil water
64 1-pint bottles (use proper beer bottles because other bottles can explode under the pressure of fermentation)

Method
1. Put the hops in a muslin bag and add them to the water.
2. Put in the malt and bring to the boil. Reduce the heat and simmer for about 20 minutes.
3. Remove the hops and add the sugar, ensuring it is dissolved, and bring back to the boil. As soon as it is boiling, turn off the heat and while it is still at a scalding temperature, pour into the dustbin and leave to cool to blood heat.
4. Add the yeast, put the lid on and let it stand In fairly warm place, no less than 60-65 degrees F for about 5-7 days or until the fermentation has stopped.
5. Decant and remove sediment from the bottom, then pour back into the dustbin and leave for a further 2 days.
6. Decant again and pour back and leave for one more day.

This takes 10 days altogether, and after that, fill at least one test bottle and leave for 12 hours. If there is not too much gas when removing the cap the rest is ready to be drunk – no, I mean bottled. It needs to be left in the bottle for a further 14 days or longer to reach its full maturity, but is it worth bottling when within 14 days you can have another batch ready?

Cookies, diehard boys, and a princess

All farmers and farm workers have to learn to be happy with their own company, because most farm jobs are solitary. Back in the 1970s my lifeline with the outside world was the letter, be it from parents or girlfriend. Carol and I had been going out with each other since we were fourteen, and I was besotted with her. While she was doing her A-levels and I was working all the hours under the sun, personal contact was exceptionally limited. My saviour was a man in a navy blue jacket. He was Joe Raey, the village postman. I saw him every day and he delivered her love letters to me wherever I was working on the farm.

Joe's round covered all the outlying farms and hamlets. He used a bicycle, and rain, hail or shine, you could set your watch by him. Despite only having a standard, red, "sit up and beg" GPO bike he was able to complete his round very efficiently. He took short cuts across fields, between farms, where there were no official paths, lifting his bike over hedgerows and fences to facilitate the most direct route. In doing so he avoided the long driveways and the need to follow the few narrow, twisting lanes that dissect the Northumbrian countryside. In the autumn, when all farms cut their hedgerows into cushion-like green walls, scattering thorn debris everywhere, he had a lot of punctures – some days as many as ten – but he could fix them within five minutes and be on his way. Joe and his bike were a well-oiled, highly effective delivery service. At various points on his round, cups of hot sweet tea were waiting for him to speed him on his way, but only after he had imparted whatever gossip was emanating from the village or from my letters, which he had read over my shoulder.

Of course, there is always some "jobsworth" with more power than common sense and a need to fix a system that isn't broken, so as to justify their over-inflated opinion of themselves. It was decided that Joe would be able to carry more mail and do it quicker if he was provided with a moped. So, with no consultation with the man on the job, a dictum arrived from On High. Joe was to have a moped and his bike was put in the shed at the back of the post office, where all the dead equipment lived.

One Monday morning in autumn, Joe set off on his brand new, bright red, shiny Honda 50 and before he had covered the half-mile to North Farm he'd had a puncture. This meant that a mechanic from Morpeth sorting office had to come out, a round trip of twenty-two miles, which resulted in a one-hour delay in the post. This happened twice more that day and of course Joe could not use his short cuts, nor could he lift his moped over gates and fences or go across fields. Gates were to be opened and closed on entry and exit. Farm machinery and livestock, which he could have skirted round if he were using his bike, would be blocking farmyards and roads. It took until six o'clock that night to finish his round, which was four and a half hours more than usual. The next day he dragged his old bike out of the shed and kept to his old schedule, but when he got back to base just after mid-day, there was another dictum, again from On High. Joe was to use his moped, or else. He was only a few years off retirement, so he obediently did what was demanded, despite the fact that the "powers that be" had to start sending a van out daily to help complete his round, as they could not have him working more than his daily hours. Who employed these people? And how dare they upset my personal postage service?

Living in the shadow of wealth has plagued me through most of my agricultural career. Behind the Nortons' house was a large, flagstone courtyard, and they used it in summer for socialising. I had to walk across this courtyard to reach my front door. Crossing to my home, through Northumberland's "chinless wonder" set, dressed in their finery, always made me feel like a serf, especially as I was often filthy dirty and, if I had been milking, probably a bit smelly. This was really me being an inverted snob, but the urge to express my sense of humour by doffing my cap, grovelling, dragging an old, war-wounded leg behind and generally being "ever so 'umble"

33

was almost irresistible. So, although I lived in a stunning setting, unless Timmy and Penny were away, I felt awkward venturing outside my front door.

One summer's day, the Officers of the Northumberland Fusiliers TA were using the formal gardens for a bash, having been on duty in Newcastle for a dignitary visiting the newly opened and architecturally acclaimed Civic Centre. They were in need of a venue for the celebrations that befitted the occasion, because their Colonel-in-Chief, HRH Princess Margaret was to be their guest of honour. Timmy Norton, being a senior officer in the TA had a large marquee erected on the lawn behind our house. It was an amazing affair with an oak floor, chandeliers, and the tables were laid for a banquet with the regimental silver and crystal. The whole thing was coordinated with a colour scheme of lemon and white. There were tall flower arrangements on all the tables and around the walls. The inside of the marquee was lined with lemon and white chiffon. It was so over-the-top that you would have thought some gay American wedding planner had been commissioned to design it.

I had been working the land that day and I was covered in a layer of thick, brown dust. On my way home I passed a large, black Rolls-Royce limousine with little Union Jacks at the front. Clearly Her Royal Highness had been given special permission to drive nearly all the way to the courtyard, risking the fragility of the large, brittle flagstones that covered the entire courtyard and its approach. The area in front of my house was filled with army officers in their dress uniforms, all drinking Pimm's or something similar. It was a very masculine affair, save for the princess, who was only just visible as a throng of excited young bucks surrounded her. At about seven o'clock the courtyard emptied and the gaggle hushed and started filing past my cottage window to the marquee, for their formal dinner. Princess Margaret was the only woman I saw and, in amongst all the dark uniforms, she stood out a bit because, if I remember correctly, she was dressed in a long, canary yellow number.

At 10.30pm I had to go out to check on a calving cow, who was in a field called the Badgers' Holes, which lay in the farthest corner of the farm, a good ten-minute tractor ride away. So having donned my wellies, I ventured out into the dark and crossed the courtyard. The marquee cast a yellow glow and a babble of sound broke the silence of the night. I was surprised to see the royal Rolls had been moved into the darkest corner of the yard and was almost blocking my route to my John Deere tractor. As I squeezed past I was amazed to see a pair of naked female feet planted firmly on the offside rear passenger window and couldn't help myself taking a closer look. It was very dark, but I could clearly make out a pair of naked male buttocks pounding away. Whoever owned the female feet was extremely vocal and I would say by the squeals of encouragement that she was really enjoying the experience. I was shocked but not so shocked that I didn't think to myself, 'the things you see when you don't have a camera!'

An hour later, after a successful calving, I saw that the Rolls had gone, and as I made my way home the hollers and hoots from the top lawn sounded more like Newcastle United Football Club's "Toon Army" of fans rather than Northumbria's elite. The Officers, so-called gentlemen, had abandoned the inside of the tent and were daring one another to climb to the top of the marquee, slide down the roof and jump into the herbaceous border. When it got to three at a time performing the gymnastics, the whole tent collapsed with an almighty crash as crystal glasses and tables smashed under the weight of the collapsing telegraph pole sized tent supports. As I helped tidy up the devastation the next day, all that was said in their defence was that it was just the boys letting their hair down. They had been under a lot of stress for the past few days, poor things! If the village boys had done anything like this, they would have been described as vandals and probably would be appearing before the local magistrate, who, come to think of it, would probably be Timmy or one of his cronies.

The Nortons had only a passing interest in the farm, but they were passionately interested in their horses and fox-hunting. A stone archway off the yard led to the stable block, a stone built horse palace with a cobbled courtyard and polished oak stable doors, which was attached to one corner to the main house. Usually between eight and ten horses were kept there, four belonging to Tim and Penny and the rest liveried – that is to say looked after by the Nortons for others, for a fee. The place smelled of new mown hay, saddle soap and sweaty horses.

Dairy cows and duck races

Nellie and Eva Cook, known collectively as "the Cookies," were employed to run the stable. They were spinster sisters and queens of this impressive facility, and it was their duty to have at least six horses fit and ready to hunt three times a week. It may be a bit misleading to call them queens, as this conjures up somebody who looks like Elizabeth or Cleopatra, but nothing could be further from the truth. Eva, the younger of the two, was about five feet, eight inches tall and five feet wide. Nellie was five feet tall and five feet wide. Both wore slippers while shuffling around their small tied cottage and wellington boots outside in the yard or fields. They were so fat that their calves did not fit down the legs of their boots, so they cut the backs out, right down to the heels, turning them into slip-ons. They wore headscarves Monday to Friday, which hid their hair rollers. Then, on Sunday, they would expose to the world the scant tufts of curly hair.

One thing you noticed about them was their teeth, or rather the shortage of them. Those that remained were black and green. They were a pair of happy witches, waddling around in their floral print pinnies, administering their lotions and potions to their charges with an ability that was legend in the county. Local vets, who were struggling with treating a horse, would arrive at the stables, collect Nellie and whisk her off in their now lopsided car to look at someone's nag. Returning an hour or so later, Nellie would take little sis on one side and five minutes later, Eva would waddle off up the track beside their cottage to the woods and hedgerows. She would collect various plants, herbs, bark and other general witchy stuff. Then they would concoct something for the vet, occasionally accompanying him when he visited his charge. Today we would call them horse whisperers, but two hundred years ago they would have been burnt at the stake.

It was Nellie and Eva that Mr Hall's wife, Joan had arranged to make my midday meal. Of course for me, a fast-growing and permanently hungry seventeen-year-old lad, Nellie and Eva's role as provider of my sustenance was their most important job. However, their idea of lunch for one was sufficient food to feed a family of six. It always started with a serving bowl of broth, thick and meaty, that would have satisfied TV's Hairy Bikers for lunch. Next there was a meat plate holding the main course. Typically about one third would be mashed potato, the next third, two veg and the last third, two homemade plate pies, piled one on top of the other. Then there was the pudding, usually jam roly-poly or spotted dick served with a two-pint milk churn of thick, yellow, homemade custard. This daily feast would be put on an over-large tray, a clean tea towel laid over the top and be put out at ten past twelve, ready for me to transport it to my cottage, which was about a hundred yards from the Cookies' cottage. As soon as I got this feast home, I got out some normal sized plates and bowls and served myself two portions of everything. I put one complete meal in the fridge for the following evening and the other I ate fresh.

The remaining food, about one third, I returned to Nellie. She scraped all the savoury leftovers into the cauldron (actually a jam pan) that was standing on the edge of the black leaded wood-fired range that filled the inglenook. This was the broth and it boiled all day every day, then on a Saturday night it was emptied out and restarted after Sunday lunch. It tasted different every day with the new additions and fresh herbs, including wild garlic, rosemary, sage, and other witchy stuff. The pan did not have a lid and you would have thought, simmering as it did, that it would have boiled dry but it didn't because it was covered by a one-inch thick layer of liquid fat. This was removed before the liquid was served as soup but was returned to the cauldron to be used on a Saturday night.

Saturdays were very important in the Cookie coven as their twin brothers arrived on the midday bus from Morpeth. As they waddled up the street from the bus stop, dressed in matching three piece suits and battered, brown felt trilbies and looking like two matching Toby jugs, they carried with them four large bottles of Johnny Walker's whisky, a large sack of lemons and two bags of Tate and Lyle sugar. As you might expect, warlocks and witches have rituals to follow and the Saturday binge was theirs. At 6 o'clock prompt, all four bottles would be opened and then they were committed to drink them because, by their rules it was grossly unlucky to take a partly consumed bottle of alcohol into the Sabbath. In the meantime the blackened kettle was put on to the range to heat up. The best glasses were retrieved from the heavily carved sideboard and lemon juice, sugar and hot water was mixed with the whisky to make hot toddies. This is where the soup fat comes into the equation; they filled four large mugs with the hot fat and left it to cool, but not long enough so that it congealed. Before they started drinking the alcohol they each drank a mug of fat, downed in one, "to line the stomach." Then the hot toddy, all of it would be consumed.

Were they drunk by the end of the evening? They were roaring. I only ever visited them once on a Saturday night and I was chased around the table by these two cackling hags, each with smudged red lips and bleary, blood shot eyes. I only just managed to wriggle out from Eva's grasp before the mountain of flesh enveloped me, from where, I'm sure, I would never have emerged. I was terrified. To me, these were two nice old ladies. Sure, they were a bit strange, but they cooked so nicely and looked out for me. At seventeen I was so naïve.

The next day I was passing through the yard to get to my tractor, which was parked in front of the Cookies' cottage. Yes, the stock still needed tending, even on a Sunday. And there they were, bright eyed and bushy tailed, in their clean pinnies and nicely quaffed hair, humming and doing little shuffling dance moves, going about their normal routine, none the worse for wear. In fact they were in such fine form you would have thought they had just been on holiday or taken those vitamins that enable you to perform "like you but on a good day."

Nellie and Eva cooked my midday meal for about three years and they educated my narrow-minded taste buds to appreciate things that would be deemed completely barbaric today, like badger ham. This is fabulous, the best ham I have ever tasted. Then there was young crow pie, made with just the breasts from fledglings taken from the nest, hedgehogs baked whole in clay (they taste like fatty chicken), squirrels, adder and snails (it's not just the French that eat them). Basically, they caught and cooked anything that moved and was big enough to handle. I was never told before the meal what I was eating but in all the time that they fed me there was only one thing I did not like and that was fox. It has an acrid taste and is very chewy.

It had never occurred to me how these two old biddies managed to get a hold of this proteinaceous bounty from the countryside, until one day I was ploughing a twenty-five acre field next to the parkland that lay behind Whalton Manor house. The park wasn't exactly Capability Brown but it was very dramatic nevertheless. Ploughing is a slow process and once you set your field out, you seem to merge with the horizon and you can watch all the animal and human movements unobserved. It's as if you and your tractor have become invisible to the animals and birds, be they deer, hares, rabbits, or stalking foxes; you see it all, while all around, a swirl of seagulls chase each other, demanding the first pickings from the newly turned earth, squabbling and squawking over the fat, glossy, juicy, worms, and the blind, pink baby mice that Mr and Mrs Tittlemouse abandoned when their world was literally turned upside down. The adult mice would skip and dance along the last furrow, trying to escape, but they would be spotted by a buzzard, who would end their terror quickly and efficiently, taking this bounty back to its eyrie to be consumed at leisure by its tufted chicks.

I had been ploughing for about an hour after lunch when I spied Eva entering the far side of the field neighbouring mine. She had a stout, ball-ended walking stick in one hand and her whippet, tugging on her lead in the other. She bumbled over to the fence and tied the pooch to a wooden rail in the shade of a large spreading oak. Then she set off around the outside edge of the field on a first circuit. With her rolling gait this took at least 20 minutes. She then did another tour of the field a few yards further in, then another, and another, the circumference of her trek diminishing with every circuit as she spiralled around the field. After two hours of watching this behaviour, I was intrigued. I thought she must have lost something the day before and was searching systematically for it, or she was picking mushrooms or some of her witchy stuff. After another hour, when she was nearly at the centre of the field, I spotted a hare sitting on his hind legs, as still as a statue. To my amazement it did not run away as Eva gradually got closer to it. Eventually she was within touching distance and suddenly, with the speed of a striking cobra, she clamped her free hand around the unsuspecting creature's ears, swung it in the air, clubbed it on the back of its head with her ball-ended stick and it was dead. If I hadn't seen it with my own eyes I would not have believed it possible.

A few days later, after I'd eaten my portion of hare and badger ham pie and was returning my tray, I asked Eva how she had done it. She just giggled and winked at me, which is a very unnerving thing to do to a seventeen-year-old. Nellie came to my rescue by explaining that there was no great mystery to this. A hare's eyes are set on either side of its head so that it has as near as possible 360 degrees of vision. So, by walking slowly around and around them you become part of their horizon and they aren't bright enough to notice that the horizon is getting bigger. If you misjudge the final sweep and snatch by a millisecond or millimetre the spell is broken and the hare escapes. When Nellie had explained the method, Eva told me her father had shown her how to do it. She had never failed to catch her tea. I am now in my early sixties and I have tried to do this and never succeeded. Who says there are no such things as witches?

I had sent for a prospectus from the Northumbrian Institute of Agriculture at Kirkley Hall. This was only five miles away from Whalton and I had the idea of doing a few part time courses there. I hadn't given up on the idea of going to Auchencrew, but the appeal of an academic qualification was waning fast. I could not see what I could gain from it when Mr Hall was teaching me so many practical skills. What did appeal to me was to achieve craftsman status, which I could gain at Kirkley Hall, as well as the obligatory wage increase that went with it. To achieve this you had to be examined in six or more different practical applications, such as milking, tractor driving, cattle handling and spraying. In fact there were hundreds of topics. Most of the tests were carried out after one or two-day courses organised by the college.

The craftsman certificate scheme was set up because, in the late 1960s, farming was one of the most dangerous occupations to be involved in. Surprisingly the agricultural work force was more at risk than the military, with the young and the old being most vulnerable. This was brought home to me when our neighbouring pig farmer lost his wife. She had been feeding a large white boar a square of Cadbury's chocolate every day since he had been a piglet. She wore a floral cotton pinnie when she was out on the farm and she kept the chocolate in her right hand pocket. One day her husband was at market so she was not doing her normal routine and in her haste she forgot the chocolate. When she reached the boar's enclosure he was waiting in his normal place and the woman went for the chocolate that wasn't there. The boar was aware which pocket the chocolate should have been in and when it wasn't forthcoming he just helped himself. He bit a large chunk out of her side and when he didn't get his chocolate he tossed her over the barrier in disgust, where she bled to death. She was discovered that evening when her husband came back from market.

Another horrendous accident happened in mid-June of the same year. I was tedding hay (fluffing up the rows of hay to let it dry out) in a field well away from North Farm and bordering a neighbour, who was baling his hay at the same time. Although I could not see clearly exactly what he was up to, I did see the tractor stop for a long time and assumed he had broken down. I later learned he had a blockage in the chamber of his bailer and had foolishly left everything running while he attempted to unblock it. His sleeve got caught and the cross auger dragged his arm out of its socket at the shoulder and into the bale chamber. This obviously stung a bit, but with just one functioning arm and a weeping stump, he stopped the tractor, retrieved his severed arm from inside the baler, and then walked a mile back home with it tucked under his good arm. Not surprisingly, as he entered the farmhouse kitchen, he collapsed into his wife's arms. The reason he had survived and not bled to death sounds a bit grizzly, but is true. Because the arm had not been cut, but pulled out of its socket, all the blood vessels and nerves were stretched until they were as thin as fine threads, which allowed the blood to coagulate quickly and seal the wound. As far as I am aware, the surgeons managed to reattach his arm and, after some considerable time, he returned to his duties.

The government was well aware of the number of vulnerable, unskilled workers there were in the industry and instigated the craftsman certificate system, with a financial reward. It quickly became the minimum qualification for farm staff. It worked, and over the next ten years the number of deaths in and around farms fell dramatically.

As for me, I achieved my craftsman status and saw life at Kirkley Hall close at hand. I decided not to go to Auchencrew to take a full time academic course, and instead followed up my new qualifications by taking a part-time City and Guilds sandwich course, also run at Kirkley Hall. This would take me the next six years to complete, but it enabled me to keep my independence, because I would have a house and an income. I would also aquire a maximum amount of practical farming experience while I was gaining formal agricultural qualifications. Mr Hall actively encouraged me to follow this avenue, so long as I made up the hours I missed on the farm when I was attending college. The courses were split up into one week every month of the academic year, and along with travelling time this added up to forty-five hours, so this was my monthly debt to the farm. It was not as onerous as it sounds, as it involved working ten hours each weekend day each side of the course and an hour extra each day. Of course there was around three hours' homework per day to fit in as well, but this just prepared me for life as a farmer. Most livestock farmers work an eleven to twelve-hour day. Even recently I spent some time as a contract milker at a farm near Reigate and a ninety-hour week was the minimum expected for me to be able to carry out the job.

The blackberry season started in earnest each year with Joan Hall scouring the hedgerows for the first fruits to be picked and, when she'd gathered enough, she would make a delicious cordial from them for our mid-morning drink. This would be boiling hot and Mr Hall would drink his down in seconds, telling me to drink up, when I couldn't even pick the mug up without scalding my fingers. After several sips of this gorgeous drink I would have to abandon it, to get back to work. Over the years I developed an asbestos lined throat that enabled me to drink all the hot cordial in the allotted time. Although my tastes have developed to strong, black espressos, the ability to finish first is deeply ingrained.

Whalton is still a very pretty village and a popular place to visit but back then, during what we called Blackberry Week, which was the autumn half term school holidays, our fields and hedgerows were ravaged by the marauding hoards from Morpeth, all of them in search of the juicy, black gold dripping off the hawthorn and brambles on the field margins. These trespassing pickers fell into three main categories. The first were young teenage groups who, when challenged, would run off at breakneck speed and disappear into the distance like Road Runner in one of Warner Brothers' cartoons, never to be seen again. The second were the retired couples who would, like the teenagers, disappear into the distance, except that they would do the same but very, very slowly. The third was the family, mostly the husband, along with his wife and two or more children. Some primeval hunter-gatherer instinct possesses the dominant male and he would stand and argue the toss about what right I had to challenge him. He wasn't trespassing; they were doing no harm (such as trampling down the hedgerow so little Emily can reach particularly juicy ones that are always just out of reach). He was The Man and no be-wellied, oiky, country bumpkin was going to tell him what to do.

After a few encounters like this I decided to change my approach from, 'You're trespassing, sling your hook!' to, 'hello! What a lovely day. I've just come over to tell you that I sprayed this field a few days ago with glycophosphate. It takes a few days to start working, so I'm sure those blackberries will be OK, but I wouldn't eat them. It's up to you. Enjoy the rest of your day. I'm sure they will be safe.'

It never failed! The man would herd his brood out of the field like some bossy cockerel with his chest puffed out, full of his own protective importance and as quickly as they could organise themselves, would drive off in the family saloon. This method of eviction had an added benefit; invariably, they would throw away the picked blackberries in a nice neat pile and of course it would have been a sin for them to go to waste.

Some of our major trespassers were men (very rarely women) desperate for a pee. Being British, standing by the side of the road was never private enough, so British men feel they have a right to invade privately owned farmland and use it as a public toilet. Of course, if we country folk peed in their gardens there would be hell to pay. One day, while I was fencing in the parkland behind the Manor House, a Mercedes sports pulled into the gateway and a Sean Connery look-alike danced into the field, clearly desperate. He opened the gate and trotted up the slope to the main body of the field, where an internal, lightweight fence brought him to a halt. He unzipped and fought to get the python out of his lair, then directed a fire hose-style stream straight onto the fence. What he failed to realise, was that the fence was electrified. Both his feet left the ground at the same time, then he collapsed to his knees and rolled over in shock as 10,000 volts coursed through his penis. Of course he lost all control of his bladder. He wet his trousers, his shirt, his shoes, his socks and he even managed to get it in his hair.

I laughed until I cried.

Cèilidhs, marriage and cows

In Whalton village hall there is a wall known as the Moor Cock Wall. It is located on the south of the building and it was named by the caller who came along with Moor Cock Cèilidh Dance Band. A caller is the person who calls out the instructions to the dancers at barn dances and cèilidhs, such as, 'now turn and face the Moor Cock wall!' Or, 'right hand basket, stopping with your back to the Moor Cock wall!' Scottish Country Dancing was a complete mystery to me, because I had only danced at discos in a Gateshead youth club, so I was particularly in need of instruction. It was incredibly energetic, and learning to keep your feet when a twenty-stone corseted rhinoceros of a farmer's wife swung you round was no mean feat. A film of it would make record viewing today on YouTube.

Carol would come up at weekends and stay over (*ooh la la!*) and we would go to most of the village dances. Virtually everybody in the village, young and old alike, danced and danced until the witching hour, and there being a distinct lack of girls in the village, with most of the dances involving changing partners, we only got to dance with one another a few times each evening. We had to overcome a lot of parental disapproval before these unchaperoned weekends occurred, and they were too infrequent.

Our solution to this was to get married. I don't recall the decision being a big romantic moment, just a natural progression. After all, we had been a couple for a long time, I was permanently employed, and most important of all, I had a house to live in. It was easy, and the only relevant objection I recall was that we were too young, because I was 19 and Carol was 20, so, amidst grumblings from parents on both sides, we became Mr and Mrs Philip and Carol Dixon. We did the whole thing: top hat, white dress (made by Carol's mother, who was an accomplished seamstress), church and then the reception at the Springfield Hotel in Gateshead. The day passed in a surreal haze and before we knew what was happening we were on a plane to Corfu. It had always been a dream destination of mine after reading Gerald Durrell's *My Family and Other Animals* during the long dark winter evenings on my own. It is a very different place now, compared to the simple agricultural backwater it was then, with its orange groves, donkeys and practices more reminiscent of Roman times than the late twentieth century.

We stayed in a tiny hotel on the edge of the Ionian Sea, ate olive oil bread and barbecued sardines and swam in the balmy sea. We even experimented with drinking *Domestica*, the local wine, which was put on our table each night. It seemed very foreign and exotic to us but the time whizzed past and all too soon we were back home.

The weekend of our return to Whalton was a real treat. The Nortons were away and the big house was empty, so we had the place to ourselves. Even the housekeeper, whose quarters were next door, was away, and we made the most of our glorious surroundings. We picnicked on the lawn, sunbathed on the terrace, and pretended we always lived like that.

As with our honeymoon, the time disappeared in a flash, and, just as we were clearing away our evening drinks from the courtyard, the housekeeper and her nine-year-old daughter appeared at our front door. She had locked herself out, and the Nortons were not due back for three more days. The doors to the Manor were built to keep out the marauding Picts from across the border, so, as you might imagine, they were quite substantial, being built of three-inch thick oak, with large metal studs embedded in them.

I armed myself with a screwdriver and a bunch of old Yale keys, but I had no chance against this level of fortification. We walked round the building twice, examining every window on the off-chance one had not been fastened properly, but we had no such luck. It looked like we were going to have a couple of guests for a few nights, but then I had a thought. Perhaps I could make my way to the main body of the house through the roof space. Taking a ladder and my cow-finding torch, I climbed into the loft space through the hatch in our bathroom, and sure enough there was no dividing wall up there. In fact, a lot of the rafters were boarded over and with loads of room under the beams, walking around was really easy. I carefully made my way forward, in case any of the wood was wormed, but it all seemed to be in excellent condition. I walked the length of the servants' quarters and got to where our wing met the main body of the house. There was enough space in there to hold the Hunt Ball, but the only drawback was that I couldn't find a hatch down into the rest of the house.

After walking up and down a few times I spotted a metal ring set into the boards, and on closer examination it appeared to be attached to a trap door. I pulled at the ring and it swung up soundlessly on well-oiled hinges, revealing a more normal looking hatchway into whatever lay below. It was a bit puzzling, because this came upwards as well, so it was also designed to be opened from above.

When I opened it all up I found I was looking into a large Victorian-style bathroom. Up till this point I felt like I was helping a neighbour in distress, but as I dropped to the floor of the bathroom and saw the feminine underclothes scattered carelessly around, I felt more like a pervert. When I opened the bathroom door I found myself in the master bedroom and it really did feel strange, but being a nosey Geordie I had a good look round at all that opulence before I made my way downstairs and opened up the door to let the housekeeper in.

Over the years the strange operation of the hatchways has had me speculating about why they were built like that. Perhaps it was used for a Lady Chatterley-type situation, or a cad returning from a clandestine rendezvous. I don't suppose the Nortons ever found out about this route to their boudoir, and of course in these more liberated times the need for clandestine routes has all but disappeared.

The next day I returned to work, and Mr Hall had some great news for me. A cottage opposite the Home Dairy had become available. Would I be interested in moving to it? I knew the house. It was a very pretty double fronted semi-detached cottage with roses round the door. I jumped at the chance. I had never enjoyed living in the courtyard, and although fronting on the main road into the village, the cottage had the welcome privacy of a walled back garden.

Most of the tied houses I've come across in all the time

I've been farming have been in a mess. This cottage was no exception, but we were supplied with as much paint and wallpaper as we wanted, so that we could make the place habitable. We drafted in Carol's parents, Eric and Violet, to help decorate and within a month had moved in.

Bizarrely, the farm's dairyman, who was our next-door neighbour, was called Tom Hall, the same as our joint manager. He was a cheerful gingernut of a man with a house full of small girls and a back garden full of budgies in wire mesh aviaries. Whenever Tom was going about his daily tasks he would whistle just like one of his budgies. This happened so frequently that the local starlings took to impersonating him, so when he was working around his cows there was a cacophony of sound as a hundred or more starlings joined in with his bird song.

On the other side of Tom's house was a smithy and, over the years we lived there, a series of men worked their trade there. All of them were strange and they confirmed the belief I had when I first met Jack Young, the blacksmith at Cambo, that they all were ruled by the moon, with moods that swung like Big Ben's pendulum. One minute they would pass a light-hearted remark, like, 'I don't know how you manage fondling all them titties of a morning. It would really upset me.' The next moment, a red hot shoe would fly out the door, missing pedestrians by inches and land hissing in the middle of a wet road. We avoided all contact with the smithies.

Our other neighbour was Geordie Nichols, a farmer. He lived alone in a house that was about a hundred and fifty yards away from ours, although a large tranche of his land was clearly visible from our back garden. He was the hand-knitted yoghurt type; he wore leather shorts summer and winter and no socks underneath his ankle boots. In fact, I think they were clogs, because, they had wooden soles with metal studs and a small horseshoe-like wear plate on the heels. You could hear him crunching his way along well before you could see him. He was a big man, with an unruly mop of black, curly hair and he was always in desperate need of a shave. He always seemed to be in bad mood, talking to himself. I found him really scary. Back then we all thought he was mad, but now he would be considered very green.

One spring he decided that it would be a good idea to grow a crop of mustard in his ten-acre field, which was directly opposite our gate. This, he said, when it was incorporated into the soil would act as green manure before he planted winter wheat the following autumn. Today, because an enormous range of crops is grown, this venture into the unknown would be no problem; the kit and machinery is available to cope with all eventualities, but back then it wasn't, and Geordie's experiment became a catastrophe.

He sowed the crop in the spring and then it grew, well – like mustard! If it had been competing against Jack's beanstalk, the mustard would have won. By the end of summer it must have been almost five feet tall, and to make things worse, secondary thickening had taken place and the plants were like small trees. Instead of trying to mow the crop, he decided, in his infinite wisdom, to roll it flat with a four-ton grassland roller. After all, the idea was to leave the crop in the field to act as manure, rather than harvest it. The trouble was, he ended up with a carpet of matted vegetation almost a foot thick. At this point he called in a local contractor, whom he instructed to rotavate the field. This process is supposed to chop up the top ten inches of soil. Unfortunately, all that happened was that the plants wrapped around the rotavator blades and clogged the machine solid. They tried putting a bigger tractor on to drive the rotavator at twice its normal operating speed, but that just resulted in mechanical things breaking.

He decided that winter wheat was not possible and that the next crop would have to be spring sown. This would allow the winter to break down his foot-thick mulch. It did succeed in allowing the leaves to drop off the remaining matt of sticks that were still attached to the soil by their root structure, but to make matters worse, this layer of debris acted as insulation. The previous year's mustard seeds survived the winter and in the spring started growing up through the carpet of sticks. The field was fast becoming a jungle – a real "no-go" area. His only answer was to spray with 'Roundup' weed killer. 'Roundup' is effective and non-polluting, but its use went against the Soil Association rules for Geordie's accreditation for organic status. The only weed killer the Soil Association allowed at that time was concentrated nicotine, which as you can imagine, is highly toxic. But, because it is produced naturally, they say it's acceptable. It's a mad world!

Anyway, Geordie sprayed the field with something. I never found out what, but it was very effective. The next summer was a hot one. The mustard sticks dried out and he was able to push them up into piles with his loader tractor and set fire to them. It took three full years to rescue the field and when he returned it to grass, because of the fires, there were funny patches all over it for years to come.

One day when I was driving down the road into Morpeth to do the weekly shopping, I passed, as I always had, a boarding kennels and working Springer spaniel dealer. I had always liked the way that Springers, when they ran through tall grass, or were just brimming with *joie de vivre* would bounce on all four legs and spring into the air, much like gambolling lambs. On an absolute whim I called in to see how much they cost, and came away an hour later with Tosca.

Tosca was a runt. He had bowed front legs, an undershot jaw, one long hairy brown ear and one short, almost bald, white ear. In the main he was while, apart from a heart-sized liver coloured spot in the centre of his back. He looked comical, but if you laughed at him he would go into a huff for days. He had all the papers to say he was pedigree spaniel, but the breeder had just given him to me because he wasn't up to their normal standard. Spaniels are loyal and obedient, but as in most things with Tosca, he only got one of these right. Wherever I went, he would follow, only allowing himself a ride on my tractor if I was on a tarmac road, otherwise he would be trotting behind me. And I mean, behind. If I had a trailer attached, he would be in between the tractor and the trailer. It was incredibly dangerous for him, but the only injury he ever sustained was that his hips wore out after a lifetime of following his master. On the other hand, if you gave him a command, he would put his head to the side and look at you, as if thinking about what you wanted him to do, or not to do. He would weigh up whether or not it was worth the ticking off, and then he would go ahead and do what he wanted to do regardless.

Marc Chisholm was our butcher. He arrived in his van twice a week at 6.30pm outside our cottage. Often there was not a large choice, because we were about the last call on his daily round and his van was un-refrigerated. However, he always swore that he had saved some sort of delicacy or other especially for us. He was a lovely little rotund man, looking a lot like one of the Home Pride flour graders, the one that wore round National Health spectacles, except instead of a suit he wore blood-stained whites. Because we were his last call for the day, he was always very generous with the price and the quantity. Presumably some of his stock was not suitable to be chilled and taken out on the next day's round. We were never ill from eating any of his meat, but we bought some strange cuts and had to be quite inventive in how we cooked them.

At Christmas this was not the case, because we pre-ordered so that we ate what we wanted and not what he had left. One Christmas Eve it got to 9pm and Marc had not arrived. We were getting concerned. Had he forgotten us? 10pm arrived and no Marc, then, at 10.30pm the sound of a tooting horn took us to the door. We opened it and were greeted by a well-sozzled butcher. He had partaken of more than one or two seasonal drinks on his round and he was clearly not fit to drive. He was swaying and giggling to himself, and in his hand, swinging by its neck with its feet trailing the ground was our turkey. Thrusting it into my hand and saying a fuzzy 'happy Christmas,' he swayed back to his van to get the rest of our order, closely followed by Tosca, hoping for some scraps to be thrown to him as Marc trimmed our meat.

Just as I was about to go out to help carry in our order, I heard a painful yell coming from the van and I rushed out to see what had happened. I was met by Marc, who was clenching his left hand. Fresh, bright red blood was oozing between his fingers and dripping onto the ground. We immediately took him inside to administer first aid and were horrified to find he had cut the end of his index finger completely off. I was about to go to the van to find the severed article when Marc told me I was too late. As he sliced his finger, he jumped back and in doing so had flicked his knife; the finger end had flown out of the back door of the van and straight into the waiting jaws of Tosca, who had swallowed it without blinking an eye. Then I suddenly realised, where was Tosca? There was the butcher's van, with open doors and nobody there to stop him helping himself. Bellowing his name, I shot outside, to find him sitting on the cutting block in the back of the van tucking into a forty-pound block of butter. I took him out and shut him into the back garden, and when Marc's wife had collected her husband, I went back to look at him. For Tosca, basically a white dog, to turn pale was quite an achievement, but he succeeded. Even his pink nose was pale. I have no idea how much butter he had wolfed down, but the upshot was a fit of vomiting that Rab C Nesbit would have been proud of, and diarrhoea that lasted three days.

Marc returned the following week with our New Year order, wearing a big bandage and feeling proud of the fact he had lost a fingertip, but he was back at work and keeping the farming folk of Northumberland well fed. As for Tosca, he hadn't learned his lesson, but that tale is for another day.

Carol had settled into country life. She was a good homemaker and my camping life-style had long since disappeared, but with me out of the house her days were long, so she found a job with an accounting firm in Newcastle as a trainee chartered accountant. Her one-hour commute to work and her home study for her exams, coupled with me rising early for milking meant at times we were like ships in the night.

One of the beauties of farming is that you work from home and when I worked late, she would join me on the tractor as I mowed or ploughed. On free weekends we would be inseparable, cherishing these rare moments together. One Sunday evening, just as dusk was approaching, we were passing a small dense coppice with a rickety barbed wire fence at the edge of the ten-acre field. It had not been disturbed for years and neglect had turned it into one of those wild corners hidden away amongst the neat order of the fields.

As we were passing its shadowy edge, we heard a playful yelp and we all froze to the spot. Tosca was the first one to move; he sank silently down on to his haunches, pricked his ears and stared fixatedly into the impermeable undergrowth. He was quivering and ready for action and there was no way a verbal 'sit and stay' would have worked. I took out the piece of baler band that I always kept in my pocket, tied one end to Tosca's collar and the other end to a post, I quickly caught up with Carol as she picked her way silently through the barbed wire, nettles and brambles. About ten yards in, the ground dropped away to reveal a small clearing of bare earth. In the middle of it were four small badger cubs, playing rough and tumble with each other, oblivious to the people peeking at them through the natural camouflage. However, mum badger, crouching in the shadows on the far side of the play area, was well aware that we were there, and was staring intently at us.

I had crossed swords with an irate badger before, when my tractor wheel had sunk into a sett in the middle of a cornfield. I had got out of the tractor cab and was standing by the big back wheel trying to fathom a way of freeing the tractor when the boar broke the surface, digging his way out of his unexpected burial.

He didn't hesitate; he flew at me, snarling, and if Tosca had not intervened, I am sure I would have been bitten. I jumped onto the steps of my tractor and a sort of Mexican standoff occurred between the two of them. Eventually, after several minutes Mr Brock turned his back on Tosca and trotted away. Tosca thought this was a signal to charge after him but the wise old creature turned and snarled. For once, my trusty companion got the message and let him go unhindered. From this experience I had learned to treat badgers with utmost caution. As long as you don't interfere in their life, they tolerate humans and just bimble along in their own world.

As we watched the badger cubs I recalled that day, so we stood back and didn't get too close while the mother badger allowed us to watch her cubs at play. That summer Carol and I often visited the set and sat and watched the cubs growing up. One evening when we were leaving to go home they followed us at a distance to see where we were going. Shortly after that the family seemed to have moved on because the site of the set became overgrown.

I was extremely fortunate with my studies at Kirkley Hall, because unlike my previous brush with 'O' levels and formal education, I found I did not have to do hours of revision prior to the exams. Once I had been lectured on some detailed aspect of calf rearing or dilution rates for sprays or anything agriculturally related, the facts just entered the old grey matter and stayed there. In the six years I was enrolled at

Kirkley Hall I passed all my exams at credit or distinction level
and I took every relevant qualification that was on offer,
some years taking one course and home tutoring myself
another. The final year I took the Stage 3 Farm Management
qualification, for which I received a distinction, and I was
the first ever part-time pupil to be awarded the annual best
student award.

But as far as student social life at the college went, I didn't
have any. I wasn't a full-time student and I had farm duties to
perform, both before and after my lectures, so there simply
was no time. That was, except for the final year ball. Because
I had won the best student prize, Carol and I were invited
to the end of year soirée, which I'm sure would be called a
prom these days.

Kirkley Hall is a 17th Century Grade II listed building,
nestled in large magnificent gardens. The main hall has
a series of French doors down the south wall allowing the
guests at an overheated summer ball to spill out onto the
cool moonlit lawns that sweep away to the distance. This
was Carol and my first formal occasion. Her mum had made
her a long dress and I wore my first dinner jacket. The full time
students had done a magical job on the decorations. They
had used the skeleton aluminium foil left over from where
the dairy stamps out the foil caps for milk bottles. They hung
hundreds and hundreds of strips from the high ceiling to head
height and then mounted spotlights to shine through the
glittering and shimmering cloud of silver above our heads.
With the group playing cover versions of Steve Wonder's
Superstition, Suzi Quatro's *Can the Can*, 10cc's *Rubber Bullets*
and (although not a very PC man now) Gary Glitter's *I'm the
Leader of the Gang*. And with all the students in their finery
(as opposed to the fact we all normally wore old clothes and
smelly wellies), it was like dancing on a movie set.

I can't put my finger on a time when I started to love looking
after cows, but the transition from me regarding them as just
being animals that were fun to be around, to my starting to
respect their intelligence and integrity must have occurred
around this time.

It certainly didn't occur because of number 75.

By law, all cows have to be numbered, and often those
numbers become synonymous with the character of cow.

For the Christian world, the number 666 is synonymous with the devil, but for me just mentioning the number 75 sends a weight plummeting to the pit of my stomach. All the naturalist types on the telly will tell you that in the wild, head and horns are the defence mechanism of the bovine species. That is true for a cow-on-cow confrontation, but a cow's main defence against its predators is a well-timed kick. When some great hairy lion is chasing a buffalo cow for its dinner it rarely tackles the cow head-on. It sneaks up from behind and pounces on the unsuspecting animal's back. When the cow falls to the ground, the lion goes for the throat, but if, when it approaches it does not manage to avoid the back legs, it will wake up hours later with a very bad head from what only can be described as a lightning-powered sledge hammer blow that can bend three-inch steel bars. The wildlife programmes rarely show this phenomenon. It doesn't make for good viewing when the humble cow drop-kicks the king of the beasts.

In the milking parlour you rarely see these kicks. Although cows may mess around when you're milking, its normally just knocks and bangs you receive. Secondly, cows rarely feel the need to be particularly aggressive, as long as you treat them with care and consideration. Thirdly, on an occasion a cow does decide to lash out, you just wake up in hospital with your head stoved in, so you still don't see it.

Number 75 was the exception.

She was evil, and every day she would attempt to kick me. Luckily she never managed to make contact, but her leg whistled passed my ears all too frequently. Putting her head down underneath the other cows, she would watch me walking up and down the parlour and if I passed too close – bang! Out came the leg and its horny hoof. Peculiarly, she was quiet while she was being milked, as long as she was approached from the side. She was not interested in knocking you about. She was only interested in the killer blow, which was worryingly premeditated. Luckily, in my career creatures like her have been few and far between.

Living so close to the dairy as we did, it became one of my duties to look after the calves, which was both a delight and a trauma. When a calf is left for its mother to rear, she does an excellent job. Unfortunately for the cow, we need her milk, so soon after birth, calf and cow are separated and the calves are taught how to drink milk out of buckets. They can be very vulnerable for the first few weeks of their lives until they get used to their new environment, and keeping these delicate calves alive is a real art. In a 100-cow dairy herd the aim is for every cow to have one calf a year to maintain a high level of milk production. Inevitably, cows do not last forever and about 25 of the cows will need replacing annually. The best 50 cows are chosen to be put in calf to a dairy bull to provide the dairy heifer calf harvest. As approximately half the calves will be female, this on average will give us our requirement of 25 purebred dairy calves. The food industry does not like beef from pure dairy breeds, because their meat tends to be dry and tasteless, and therefore the remaining 50 cows are put to a bull of a beef breed. This will give us about 25 heifer calves, which will give a good enough milk yield, about 25 bull calves, which are fattened up for beef and 50 crossbred calves, which will give the butchers a much better carcass than a pure dairy breed would.

We had previously been using Hereford bulls, which threw calves with big heads, narrow shoulders and large rumps. In the main, this resulted in reasonably straightforward calving, that very rarely needed human intervention, but not as easy as the calves produced by the Aberdeen Angus bulls that the heifers were normally serviced with for their first pregnancy. These calves had a real zest for life and often would be blaring for milk when they were only half way out of their mum.

Earlier, I mentioned Mr Leon, the little weasel of a man employed by the estate agents Campbells Green as land agent at North Farm. In my many years in agriculture I have either come across, or have had the misfortune to work with a number of these specimens. They are scheming, conniving bullies and complete leeches on society. They have never run a business of their own and when one they are overseeing goes wrong, it is never their fault! Of course it isn't...

Leon instructed us that we should use a Charolais bull on our dairy cows, because he had heard that crossbred calves from this huge French breed were making a good price in the market. We warned him it would not work, and never had worked. It would be disastrous, but Leon knew it all and the ramifications of that decision did not interest him.

Nine months later, the farm team was picking up the

pieces. Every calving needed a lot of assistance, probably one third of them the vet had to assist and over a quarter were either born dead or died during calving. It was the disaster we had predicted. We had a lovely cow called Poppy who would insist on you scratching her ears whenever she was in the vicinity. She wasn't an enormous yielder but she was consistent, was never ill and had a calf every year. She was a patient member of the herd and I liked her a lot. I was very nervous for her as her calving time came around and I would visit her day and night every few hours to ensure her survival.

One Saturday, I was on duty by myself, as both Mr Hall and Tom the dairyman were away for the weekend. We were strip grazing Jenny Wets (our fields had some great names), which involved holding the herd back with an electric fence and only allowing them a strip of about thirty yards of new grass at a time. It was my 10pm visit to the herd and armed with my torch, I made my way across the loafing (resting) field to very sleepy kine all curled up and grunting peacefully. It took about half an hour to go round them all, 'gannin' canny' and 'divn't dunch inta owt,' (meaning, "taking care" and "don't bump into anything") as Tom would advise.

Poppy wasn't there. Cows often take themselves off somewhere quiet to calve away from the herd, but knowing that and finding them are two entirely different things. I searched high and low for her and eventually found her by the straining noises she was making as the contractions gripped. She had slipped under the electric fence and made her way to the far end of the field, crossing a boggy patch, and ended up spread-eagled under the branches of a large oak tree. When I got to her, the water bag was still intact and the calf's nose and feet were nicely presented. The trouble was, they were enormous.

I looked around for a good stout branch I could use as a handle to help me ease it out, and found a piece of oak that looked right. I made sure the water bag had burst, pulled out the baler band that was ever present in my pocket, made slipknots in each end and attached them around the calf's hocks. I tied the loose ends of the string around the stave and then, as she pushed, I pulled and, little by little, eased the calf out. After ten minutes I had exposed the head and realised the sheer size of the monstrous calf I was dealing with. I pulled and pulled, but the poor thing's shoulders were trapped fast behind its mother's pelvis. I have small hands for a farmer, but even I could not squeeze them down the side of the calf to stretch the skin a bit further. It was getting desperate. The calf's chest was squeezed into the birth canal, and because the umbilical oxygenated blood supply was severely restricted, its tongue was swollen and was starting to turn blue. Time was running out. Then I remembered a trick Mr Hall had used in the past. I undid the string on one leg and pulled on the other. When it had moved a few inches I swopped over to the other and gradually managed to walk the calf's shoulders past Poppy's pelvic circle.

It was at this point Carol arrived. I had been missing for two and half hours and she had come looking for me. Although she had thrown herself enthusiastically into country life, she was not keen on traipsing around fields on her own in the pitch black. She found me stripped to the waist, flat on my back amongst the blood and amniotic fluid. I had my feet planted firmly on Poppy's pin bones and I was pulling with all my might and main, sweat almost squirting out of my pores with the effort I was exerting. Then, all of a sudden the shoulders slid out and between us we got the chest out. With a normal-size calf, when its chest gets past the mother's pelvic circle, the soft tissue of the calf's belly squeezes in hard to clear its lungs of fluid. At the final stage a big calf can shoot out the amniotic fluid like a geyser, the squeeze is so dramatic. At this point you only have a brief window of time to finish the birth, because there is no oxygen available from the mother with the umbilical cord stretched and squeezed close against the wall of the birth canal. With a big meaty calf you can speed the final process up by using the length of the calf to lever out the hips by pushing down with its body. With one last tug it was out and I dragged the beast around to his mum's head for her to start cleaning it up. It was alive, but only just.

Poppy was flat out, away with the fairies and showing signs of milk fever, a condition caused by a deficiency of calcium. She needed a bottle of calcium borogluconate. I left Carol rubbing the calf dry with my coat and shirt in an attempt to get its circulation going and I ran up to the steading for a flutter valve and a bottle for the cow. When I returned with the necessary, Poppy was still flat out, virtually unconscious. I stuck a large needle into her jugular and connected the long rubber flutter valve to the warm 500cc bottle of maximum strength calcium borogluconate, making sure no air got in, and watched the level in the bottle go down. This takes a matter of minutes and the result is often dramatic. A quarter of an hour later she was on her feet, seeing to her calf's needs as best she could.

It would have been nice if the story ended there, with mother and calf doing well and going on to have a healthy life, but it didn't. The next day when we went to bring the calf in it was clear with all that pulling I had damaged its ankles and it was incapable of straightening them.

It is very important that a calf drinks its mother's colostrum, the "first milk," as it contains seven important elements for its survival, and because this chap could not walk, it could not suck his mum's udders. Although Poppy was a very docile animal, she would not allow me to milk her by hand in the field, and so for her comfort, her overflowing udders and the calf's needs we brought her off the field and onto the concrete apron in front of the milking parlour. There, an old boss cow butted her out of the way and she collapsed onto the floor. As she tried to get up, the tendons that hold her back legs in the upright position gave way, her legs splayed out sideways and she fell heavily onto her swollen udders. The protracted and rough calving had weakened the tendon attachment and she had to be taken to soft ground as quickly as possible before irreparable damage was caused.

Unfortunately, back in the 1970s, to do this we had to lay a gate alongside the cow, roll her on to it and tie her down and then drag her into a field. This sort of manhandling seems to do something detrimental to a cow's psyche and probably only one in twenty ever manage to get up again. They just give up. Poppy gave up and a few days later the hunt took her carcass away for the hounds. Her bull calf lasted about three weeks. He never managed to straighten his legs and although he had another cow's colostrum, he contracted pneumonia and scour (extreme diarrhoea) and he never gained a zest for life.

So, Leon's decision for a slight increase in calf prices cost the farm dearly and, of course, he made sure the responsibility for its failure was allocated to the ineptitude of the farm staff, with no hint of blame to blemish his impeccable reputation.

It was during this period of bad calvings that I helped at my first cow caesarean, which are conducted in very different conditions from a human one. Mr Hall and I were trying to calve a cow in a box next to the dairy. He was the boss of course, so was taking the lead. I was just holding the cow on a halter tethered to a ring on the wall, trying to keep her still, while he did an internal examination. After a few minutes, he asked me to have a go and tell him what I could feel. I have always had small hands so I assumed he was struggling to get his shovel-sized hands and forearms in far enough past the calf to see what was going on and what was holding the job up. Was it another enormous calf? Her pin bones (part of the bony structure surrounding the rear end) had dropped further than I had seen before. There was enough room for a coach and four to get through. I stripped off my shirt and slid my arm inside, avoiding slipping and falling in, never to be seen again. All I could find with my fingertips at the fullest extent of my arm and my ear resting on her backside was some fur covered skin. After a quick conflab it was decided to call the vet, as Mr Hall had never come across a calf lying so deeply in the body of the cow, and of course if he didn't know, I certainly didn't have a clue.

Bruce Robinson, the senior vet in a large local practice arrived forty-five minutes later. He was probably the most straight-talking vet I have ever had the pleasure of working with. If the problem were beyond his ken, he would say so and always come up with some innovative solution or treatment, which allowed you to have confidence in his decisions. When he performed his internal examination on the cow, he had the same problem as Mr Hall had; he couldn't reach with his big hands and well-muscled forearms.

I explained what I had felt, to which he said, 'well, it's got to come out somehow and if the little bleeder won't come out the hole God made for her, I'll have to make another to help the job along.'

With that, he went back to his ancient Land Rover and pulled out a heavy-duty stainless steel fish kettle, which contained all he needed to perform a caesarean. We set it on a clean bale of straw in the box, together with two buckets of scalding hot water and an extension lead. He then organised us. He put Mr Hall in charge of the halter to keep the cow under control and I was to assist him. Out of the kettle came a clean cotton bag that contained sparklingly clean electric clippers. He plugged them in and shaved most of her left side behind the rib cage. I then asked how we were going to do this and to my amazement he told me that the cow was to be left standing, and the procedure would be done under local anaesthetic. With that, for the second time that evening, off came my shirt and the pair of us scrubbed the whole of our upper bodies in hot water laced with iodine, which smarted as it got into all the nicks and grazes I would gain on a farm on a daily basis. Bruce instructed me to stand back, and under no circumstances was I to touch anything until I was instructed to.

The operation area was scrubbed down with concentrated iodine and then Lignocain (a local anaesthetic) was injected every few inches down the line of incision. Picking up a scalpel, he made a large vertical incision, cutting through each layer of tissue one at a time until he opened up a hole into the womb. Normally the calf would be lying there just ready to lift out, but this was not a normal operation. The vet dived in with both hands and struggled to pull out a football-sized ball of fur, and then under that a complete but tiny dead calf, still in its water bag. For a moment, we were all speechless. Then Bruce gelled us back into action as he continued to extract the afterbirth. The whole operation had thus far taken less than five minutes.

At this point he called for me and told me to hold both sides of the womb wall exposed so he could stitch it back up with dissolvable catgut. This sounded fairly straightforward to me, but I had not realised the strength that would be needed to hold back one of nature's amazing phenomena, the immediate shrinking of the womb from the size of a 112lb sack to that of a small sandwich bag in the space of fifteen minutes. By the time he had finished stitching, my hands and arms were screaming for relief, my teeth were clenched and it felt like my eyeballs were completely out of their sockets and rolling around on my cheeks. When he finally said I could let go, I fell to my knees, too exhausted to hear their jibing remarks about being a lightweight with the stamina of a butterfly. Bruce continued stitching up, layer-by-layer, squirting antibiotic powder between the tissues and finally closing the leathery skin layer, leaving a small hole at the bottom for the wound to drain out. He took the entire foetus and the strange fur football away, to be sent to Newcastle University Veterinary research lab for them to forensically dissect. After a few days the cow re-entered the herd and after a few weeks was milking well and was even back in calf. As for the lab results, the fur ball was exactly that; a fatty ball covered with skin and fur, an abnormality I have never seen or heard of since. I believe it is still stewed in formaldehyde and on display as one of the curiosities at the Veterinary College in Newcastle University.

5

A holiday and a pregnancy

The year I finally finished all my studies, Carol and I arranged to go on holiday with Anne and Albert Brown, our closest friends. Anne was Carol's best friend from grammar school, but bizarrely Anne had been mine in junior school. She had been a real tomboy and we used to play together constantly. That was, until she started getting interested in me as a boy as opposed to a playmate. This came about when Mr Martin, the head master at Harlow Green Junior School in Gateshead, insisted that the final-year children should be taught how to waltz. He made all the boys stand on one side of the hall and all the girls on the other, then knowing that no ten-year-old boy would willingly choose a girl to dance with, he asked the girls to choose a partner. This was the breakdown of our relationship, because she always ran over and chose me. I was a boy, and girls were, well, girls! They were definitely NOT boys, so how could I possibly be friends with one? It wasn't until Carol reintroduced her into my life that the opportunity to renew our friendship occurred. Along the way she married Albert (Al, as we called him) and although he was "canny crack" (good to talk to), I was always more comfortable in Anne's company than his.

We decided a boating holiday would be the thing, and, to keep it affordable, a late season booking was the order of the day. After several late night dinner parties we decided on a boat yard at Stourport on Severn, just south of Kidderminster. We booked a traditional narrow boat for the last two weeks of September, which took up the whole of my holiday allowance. Mr Hall was happy with this because normally it was a quiet time on the farm. When we booked it, we had five whole months to wait before our break together and by the time our departure date arrived we were like four excited three-year-olds. We had paid all the necessaries up front by post and were told not to arrive at the boat yard before 2pm. I was driving and therefore it was up to me to plan the route and with my usual precision planning we arrived two hours early, only to find the large solid steel gates to the boat yard all padlocked and chained. Worryingly, the place looked deserted and foreboding. A six-foot high corrugated iron fence with barbed wire coiled along the top not helping to dispel the effect. Judging by the letters sticking out of the big green post box that hung on the gate, the mail had not been collected for several weeks. We all looked at each other with dismay.

Ever the optimist, I said, 'Well, he did say "not before 2 o'clock."' So we sat in the car, trying not to discuss what to do if nobody turned up. Anne was a policewoman, so we talked about whether her professional skills might be useful to help us find a solution. We talked about really adult solutions, like just climbing over the fence. Anne said she thought this would be trespass and would probably be a civil matter. Another useful suggestion was to hot-wire the boat and just drive it away, but Anne said this would definitely be a criminal offence. The police can be such spoilsports at times! Fortunately, hot-wiring a boat was out of the question, as none of us knew the first thing about boats. Fortunately for us, there was a three hour tutorial on boat knowledge as part of the hire agreement.

In the end we just sat and waited until 2pm, when we were relieved to see a man sweep up to the gate in a beat up old grey Austin A55 pickup truck. It must have had next-to-no steering, because there was an enormous engine strapped into the back, the weight of which had almost lifted the front wheels off the ground. A bearded man in greasy blue overalls unwound himself from the driver's seat and stood up. And then up; and then up some more. He must have been nearly seven feet tall, with the biggest feet I have ever seen, and with his massive beard could well have stood in line with ZZ Top.

He introduced himself as Ruben, the owner, and in a broad Yorkshire accent said 'Yu'll be t'Dixons. Tha's bin a bit of a change. Yon long'in bust a cylinder heed.'

Broadly translated, he was telling us that the engine in the narrow boat we had hired had bust a cylinder head and the gasket was no longer available, so he couldn't fix it. He did have a boat he had just finished building for a customer who had welched on the deal, so he was going to add it to the hire fleet. Although it was far from traditional, he said we would 'appreciate its luxury.'

He threw the big gates open wide, the post spilling out of the overstuffed box and blowing to the four winds. What it revealed was a scrap yard full of bits of boats. It did look organised, with neat rows of similar looking bits, but there was no getting away from it. It was a scrap yard. Ruben told us to park in between two enormous skips so tall you couldn't see what was inside, and to grab our belongings and follow him down to the jetty. We were saying nothing to one another, but by this time we all were in fear and trepidation of what kind of hulk that we would find waiting for us at the jetty.

We started our five hundred yard trek through the mountains of rusting steel and twisted, splintered fibreglass and, through a gap in the Stalag Luft 2 style fencing: all that was missing were the guard towers. Up to this point, we hadn't been able to see the river, but as we rounded a shed that looked like a collapsing aircraft hangar, we saw before us the glistening River Stour, meandering its way through enormous, trailing willow trees, with a patchwork of green spreading each side of it. We were utterly amazed. Constable would have been proud to paint the scene. There was an old stone-built marina set into the bank of the river, empty except for one small ocean going cruiser that would have looked good on a James Bond film set. All it needed was some bikini-clad girls draped over it.

Ruben explained that we were his last clients for the season and he had just craned his entire hire fleet out of the water for the winter maintenance, except for the narrow boat we had booked and this new fibreglass palace. As Ruben had told us, the engine on the barge had given up the ghost, but fortunately for us, this luxury cruiser would be its replacement. She was called *Sunrise*, but there were some restrictions we would have to put up with. First, her beam was too big for the local canal network, and although she had a retractable keel, she needed more depth of water than a narrow boat. This meant our holiday was restricted to the Rivers Stour and Severn. The 150-horsepower engine was powerful enough for water skiing, but (unsportingly, I thought) he had governed it down to meet the 4mph speed limit that was in place on the rivers. He did tell us he had left a lot more power than we'd need for the canals because of the "bore," which we would experience three times when we were out, and one of those times it was forecast to be quite big. We had no idea what he was talking about and told him so.

Ruben breathed in deeply and then let it out slowly through the gap in his front teeth, making a sort of whistling kettle sound. He explained about the Severn being tidal. At certain times of the year when the moon was at its highest, as the tide went out there was a wave of water that coursed its way down the river, resulting in very strong currents. That was the "bore." In our ignorance, we thought it didn't sound as if it was anything to worry about.

We made our way down to our exclusive holiday transport, a two-deck affair with a sun bridge mounted on top of the main cabin, which housed another bridge for use in bad weather. You could easily imagine fishing for marlin in the Caribbean, standing on the highly-polished teak decks. Her berths were more like stulerooms with en-suites. The galley was better appointed than my kitchen at home, and the best word to describe the salon was "sumptuous." After delivering four solid hours of instruction, our patient giant waved us off with, I must say, a face that looked surprisingly unworried, considering he was leaving four young ignorami in charge of such a superior vessel. As he threw us our bowline, he happened to mention that this was her maiden voyage.

By the time we got going it was six o'clock, so we decided we would cruise downstream for an hour and moor at a riverside pub for the night, and as we rounded a sharp bend in the river, there before us was a glorious looking pub.

Some years prior to this trip I had been on holiday to the Norfolk Broads, so I had some limited experience to back up my masculine bravado about navigating a boat, but where the water on the Broads was millpond-still, here, it was moving. We soon realised we had to go quicker than the current to be able to steer. Keeping in a straight line was an art in itself, but mooring for the night was a whole new ball game.

I pulled over to the jetty straight away and put the boat into reverse. As soon as we had slowed to less than the speed of the current we were swept sideways back into the middle, and, as the propeller was in reverse, it drove the boat backwards towards the opposite bank.

You have to be calm in these circumstances, but we panicked, and screamed at one another. We all had different ideas on what to do, but I put it back into forward gear, and in doing so narrowly missed an overhanging oak tree. By this time we were about three hundred yards past our destination, but this put us against the current, which gave us control. Panic over, and we managed to moor for the night. We were all so hyped by the day's events we voted for an evening meal prepared for us at the pub, which meant that our mooring fees would be included in the price of the meal. We ate like kings and queens, entirely befitting our new mode of transport and then, by starlight, (because no one had remembered to bring the torches, rather than for romantic reasons) we swayed our way back to *Sunrise*, only to discover we couldn't find the light switches. So, with much fumbling and giggling in the pitch black, we tumbled into our beds, not finding our pyjamas or cleaning our teeth.

The new day dawned, and by a process of elimination we found the light switches, only to discover they didn't work. Albert, despite having a name that conjures up an image of somebody who is completely cack-handed, was quite a handy chap. He was a chemical engineer by profession and with his training and my practical skills we managed to find the fuse box. Using various kitchen knives as tools, we checked all the connections and found that all was fine. We were just about to phone Ruben when I decided to try the ignition switch for the engine. Bingo! *Sunrise* lit up like the showboat on Blackpool's Pleasure Beach, because all the switches we had tried the previous night had been left in the "on" position. We left it all lit up while we had leisurely cooked breakfast, but when we then came to start the engine for the day's cruising, we couldn't, because our illuminations had drained the batteries. We phoned Ruben from a phone box just down the road from the pub and he told us to link up to the mains socket at the mooring for an hour until there was enough life in the battery to start the engine, and then spend the day cruising. By then the batteries would have taken on a full charge.

That night we stopped in the middle of nowhere, much to the annoyance of the fishermen who were every few yards along the length of the river and were always grumpy with us. As the night came in, we switched the saloon lights on and to our dismay, there was a fizzing from the ignition switch and we were plunged into darkness again. This time, Ruben sent his marine engineer out to rectify the problem. Where Ruben was a giant of a man, this chap was small, wiry and misshapen, like one of Gringott's goblins in the *Harry Potter* films. This made him ideally suited for squirming into all the little spaces around the boat. It took two hours for him to fix the problem and, with a bit of tinkering, give us the refinement of not having to turn the ignition on when we used the lights.

We were cruising along an uninteresting bit of countryside when Anne decided she would try her en-suite shower, and by her dulcet tones she was clearly enjoying it. After a few minutes the singing came to an abrupt end, turning into cry of dismay. Carol and Albert ran below to see what the problem was (I stayed up top – somebody had to steer). To their dismay, they found the floors swimming with water. On closer inspection, it turned out that the shower tray was missing a connection to the waste pipe. We moored, cleaned up the floodwater and inspected the other tray. This seemed to be correctly installed and proved to be so when the rest of us showered that night. After our showers we were all feeling a bit chilly as the night cut in and with the cold wet carpet underfoot we decided to push the button for the central heating. It was heating up nicely, giving off the smell of hot air blowing over new paint. This slowly turned to a distinct smell of ozone, and then all the lights went out. We phoned Ruben the next day, having spotted a red phone box in the far distance from our crow's nest-style upper deck.

We arranged to meet a few miles on at Holt Lock, where he could get to us with his truck full of tools. At Holt Fleet the river narrows and the current speeds up before dropping about ten feet over the weir. At the approach to the lock, the river is divided by an island and comes back together again just after the mighty Severn plummets over the weir in a cauldron of eddies and strong currents. The current in the approach to the weir, which is on the port side, is quite fast, but the water in the entrance to the lock is still. As you approach the narrow entrance to the lock you have to keep the speed up to maintain steering in the strong current that flows over the weir. As soon as you enter the lock basin, you throw the boat's engine into reverse to avoid crashing into the lock gates. This was tricky for us, as it was only the second lock we'd tackled on the holiday (actually we weren't calling it a holiday by then – it didn't seem quite right, so we had renamed it our summer adventure). Albert was at the helm and as we approached, he pulled on the gear lever to put the boat into reverse, to slow it down. Then he tried to put it into neutral as he approached the lock gates, but nothing happened. No matter how Albert moved the controls the boat stayed flat out in reverse and she shot back out into the strong current. The steering on a boat is dependent on forward movement, so the engine was sending us at full speed, with no steering, sideways across the river and towards the central island, but the river current was pushing us quicker towards the weir and to what seemed our doom.

I was at the stern, preparing to jump off to moor up when all this kicked off and, fortunately as it turned out, there had not been time for me to move anywhere else. The banks of the island ahead of me were littered with fishermen who I shouted at for help and to catch our mooring rope, but all I got were curses for disturbing their lines. By this time the blue-uniformed lock keeper was dancing from foot to foot, being about as helpful as an alcoholic in a brewery and swearing and cursing at us for being useless amateurs. The seriousness of our situation began to really sink in as we were being swept past the island and ever closer to catastrophe. I had no choice: with the rope in my hand I took a flying leap and landed on the point of the island near the weir. I managed to loop the thick blue cord around a stout tree stump, pulled hard and braced my feet against the rough bark of the capstan-shaped stump. The boat swung pendulum style in front of me and the rope cut deeply into my hands as I hung on for grim death. The prow dangerously scraped the wall of the weir, but it came to a standstill. Disaster had been avoided.

Just as we were taking a second breath, Anne's head appeared up the companionway. She had been on the loo and so had been totally unaware what had just occurred. She went pale as she saw what position the boat was in and got very confused as all three of us simultaneously began babbling an account of what had just happened.

Minutes later, a red-faced lock keeper arrived to give us a thorough dressing down, but when we explained what had gone wrong, he realised it was not our fault and he calmed right down and became quite a sweetie. We were about half way through manually hauling our stricken boat across to the locks pound when Ruben turned up. The keeper bollocked him up hill and down dale and he just stood there, shoulders hunched like a sulky teenager, looking completely dejected. The lock keeper, who turned out to be our saviour in navy blue, made Ruben fix everything that was wrong on the boat and found a few more faults we weren't aware of. Then the pair of them put the boat through its paces.

After five hours' work, *Sunrise* was declared river-worthy and we set off again on our adventure. We made our way southwards towards the estuary, which was going to involve many days cruising and the negotiation of numerous locks. For the next six days nothing major went wrong, but one at a time almost everything electrical ceased to function; shower pumps, lights, the heating, the oven – the lot. Some of it we managed to fix, but we didn't bother to report it to Ruben because it would have just held us up for another half-day.

Two Severn bores occurred without affecting us, because we had been moored in marinas when they had passed, but each had left the banks covered in six inches of silty mud. By the time we got to Tewkesbury we were getting desperate for a hot bath and to have a hot meal prepared for us, as this was becoming less of an adventure and more of an endurance course. We moored the boat at a jetty outside some very pretty Tudor buildings and made our way to the municipal swimming baths. The swimming pool was closed to the public, but they had a newly opened sauna suite and there was a mixed session we could join. What sort of Nordic decadence was this? We had all heard of saunas, with Scandinavians rolling naked in the snow and beating each other with birch twigs. We were such northern prudes, but the receptionist assured us that everybody would be very "correct" and would be wearing swimming costumes. We thought, 'why not?' So we did, and after we had changed into our bathing costumes, we all met outside the Norwegian log cabin that was the sauna. On its outside wall was a notice, which we conscientiously read, but it only told us that we entered at our own risk.

Timidly we opened the door and we were hit by a blast of hot air that might have come straight from Death Valley, except that it carried a curious menthol smell. In the middle of this glorified wooden garden shed was an electrically heated, rock filled brazier. Around the edge were pine benches. There were no other people in there, so we all piled in, giggling. We tried to sit down, but the wooden slats of the seats appeared to be super-heated and they burnt our bums. We retrieved our towels so that we had something comfortable to sit on and we all settled down, but very quickly we started to melt in the super-heated atmosphere. Beside the brazier was a wooden tub full of water, and some bright spark suggested we should cool the temperature by pouring water onto the hot rocks. This, of course, had the opposite effect. The temperature practically doubled and we all tumbled out to cool down in the showers. Then we all piled back in again for a second go, and much to our embarrassment, rivers of mud coloured sweat ran down our bodies as the heat emptied our deepest of pores of what seemed like years of ingrained dirt. How on earth have saunas developed this sexual reputation? The last thing on my mind as I sat in a puddle of dirty of sweat, feeling like a limp lettuce leaf, was sex!

When we decided we were all clean enough, satisfying our hunger became top of the list, and as we were leaving the baths I asked the receptionist to recommend somewhere to eat. She suggested a spit and sawdust spot round the corner that did a great beef cobbler and a pint of the local brew. That would do us very well, and we finished that off with massive portions of apple crumble and steaming rich yellow custard, clearly home made from scratch with eggs and cream. No cheating with Bird's custard powder here!

Arriving back at the boat, clean and well fed, we were all ready to face the return journey back to Stourbridge. Five days of our holiday had already passed and it was time to forge our way back up the Severn, because Ruben warned us it would take far longer against the current to get back to the boat yard. He was right. We'd just passed a difficult section of river, about six miles outside of Tewkesbury, and in the middle of nowhere where we had to follow a navigation channel between the rocks, when the boat's engine died. After our previous experience, we threw the anchor over board, so at least we were safe.

This was fast becoming like some sort of management training course, teaching us to solve problems by using our initiative. So that's how we approached solving it.

First, identify the problem. Easy. Four people are marooned on a boat in the middle of the Severn River. In four hours' time the biggest bore of the season is about to arrive. How do they avoid disaster?

Next, identify the resources to hand. Can one of us swim? Yes, but the river was very fast-flowing with all sorts of debris, such as logs floating along in the strong current. So this was ruled out as an option. Could we make a raft? No: short of breaking up the boat we were on, we had nothing to use.

Could we tie all the mooring ropes together, attach the life-saving ring to them, stand on the high roof of the boat and hoopla an old tree stump that was sticking out from the nearest bank? Yes, and that was the course of action we took, except that it took us an hour and a half and five million attempts before Anne successfully landed it over the aging bollard.

We gingerly weighed anchor and Albert and I reeled in the ropes, pulling us, inch by inch towards the bank. As soon as we were close enough, Anne took a run and a flying leap towards the bank. Being the fit policewoman she was, she performed a leap that was of Olympic standard, and when she landed, stopped as solid as a rock. That was not because she had perfect balance, but because she sank waist deep in river silt and couldn't move at all. Luckily she could reach the towrope that was still firmly attached to the mooring post and she managed to drag herself onto the surface of the foul-smelling black mud. When she finally stood safely on the bank she looked like she had been for a mud bath. Being the stalwart she was, she dragged the boat close enough to the bank for us all to disembark. We tied the boat up as best we could and then continued to climb up the bank to the wood at the top. The wood turned out only to be a few yards wide and to our dismay we found ourselves looking at another expanse of water. We had managed to escape onto an island.

We trudged along its length, through a tangle of brambles and six-foot high nettles dense enough to stop a herd of wildebeest. At the far end, we found a tree that had fallen across a narrow gorge, giving us a rough and slender bridge to cross to the far bank and terra firma. Twenty feet below our bridge, the river was a foaming torrent, but we were beyond caring and we just clambered on and walked across it as if it was six feet wide and not the six inches it actually was. After crossing half a dozen fields we came across a road, then half a mile along it we found a tiny village post office, where we phoned Ruben. He came and collected us, took us back to the yard while his small assistant took an inflatable down the river to collect our belongings. It felt like civilisation again and the end of our adventure.

We stayed overnight in a pub and the next day collected our stuff from the boat yard. Ruben was full of apologies, but didn't offer us a refund. He did offer us a free fortnight the following year, which – surprise, surprise – we did not take up. I asked him how he was going to get *Sunrise* home, but he said he wouldn't bother. He would let the winter do its worst and then claim the loss on his insurance. The cynic in me wonders if that had been his plan all along.

Back home, late summer turned into early autumn and I had attached my three-furrow plough to my John Deere tractor for the annual ploughing marathon. We always tried to get the earth turned over before Christmas to allow the winter cold to perform its magic, killing all the weed roots and breaking the sod into crumbs. We call this frost mould. Most evenings I would plough until darkness fell, then I would make my way home and Carol and I would eat our evening meal together. This particular evening when I arrived home there was a glorious smell coming from the kitchen and there were lighted candles and a bottle of wine on the dining room table. My first thought was, 'Oh, Shit! What have I forgotten?' But Carol said she had some news for me. She told me she was pregnant.

I was thrilled and we talked until late into the night about new plans. When the baby was born, we would lose Carol's salary, but I was now fully qualified and could look for a herd manager's job. It was all so exciting, and to add to my delight, the following afternoon as I was changing fields with my plough, I witnessed nature's beauty in a once in a lifetime display.

It happened on one of those glorious evenings just as the sun was sinking into the horizon, with the backdrop of an azure sky, a yellow stubble field, the trees and the hedgerows a deep, russet red. And in the air, thousands upon thousands of silver threads were dancing and drifting on the balmy breeze. I stood awe struck: it was gossamer. A new batch of hatchling spiders had climbed to a high point to shoot a length of spider silk, waiting for it to be caught by the wind and distribute the tiny creatures across the fields. I had seen this on a small scale before, but this display was of epic proportions. The eggs in every spider's nest must all have hatched at the same time, releasing a tsunami of gossamer, which I was privileged to witness. As the sun sank and the aerial display diminished, the sun's rays picked out the shimmering, silver carpet that covered everything. It was glorious. But I was there to plough, and so, as the silver rain fell, I set my three-furrow Ransome into the ground and it turned up the rich, brown soil like a dark wave on a calm sea.

As the sun sank below the horizon, an enormous full moon took over the job of lighting my way and as the dark cut in further, the dull, red glow of my tractor exhaust, sticking out of the bonnet, added to my memory of this unearthly scene. All I needed was Pink Floyd playing *Dark side of the Moon* to complete the picture.

Ashington Farm

On March 10, 1975 our life changed forever, when the first of my enormous family was born. She arrived in the Princess Mary hospital in Newcastle. This made her a proper Geordie, unlike her father who was born in Gosforth, which is in the metropolitan borough of Newcastle but not the banks of the Tyne. Gosforth is the posh, *nouveau riche* end, and as Alan Plater said in his play, *Close the Coal House Door*, 'they're all "fur coats and no knickers."'

I missed the birth because I, a mere twenty-two-year-old, allowed myself to be bullied by some officious midwife. 'She'll be hours yet,' she told me. 'Go home and get a good night's sleep.' I left the hospital and sat in the car for about half an hour before something told me to ring up. When I did, I was told I had a healthy baby girl, who had been born ten minutes earlier. Carol was one of the lucky ones who gave birth more easily than most, as you will hear later on. We called our daughter Sharron, and she was a delight.

Having the responsibility of looking after a child triggered the hunter-gatherer instincts in me. It was time to move on, so I began to look in earnest for a better paying job, hopefully finding a position that gave me sole responsibility for a herd of cows. I found my career move through an advertisement in the Morpeth Gazette. The job was at Ashington Farm, Northumberland, twelve miles away from where we were and it was to manage a herd of cows for the Douglas Brothers. I was armed and ready for the challenge, with a boatload of qualifications, an internal drive like a chieftain tank, a confirmed love of dairy cows and my lovely family. The interview for the post of Herd Manager was surprisingly short. From what I could tell I was the only interviewee and after I had been shown around I was surprised to be asked if I wanted the job. It felt as if I had interviewed them and not the other way round and that felt bizarre, but I still took the job.

The Douglasses, Bruce and Malcolm, provided us with a newly-renovated, end of terrace cottage bordering the farm's bull paddock. The job involved managing a herd of 125 British Friesians. Although I had a lot of experience with the Whalton herd, I had not been dealing with the whole herd on a daily basis, so I did not know all the cows individually. At Ashington Farm, over the following months, I got to know this group of females – their hierarchy, sibling rivalry, who was most hormonal – very well.

All cows and bulls have individual characters and, just like us, different levels of intelligence. Take Frieda, for instance. She was a red-and-white Friesian cow (not all Friesians are black and white), stout and quiet-tempered, and had the number 52 freeze-branded on her rump. And she was greedy. She would eat her ration, then push less dominant cows out of the way to try and satisfy her insatiable appetite.

When grazing the kine, we keep them to whatever part of the field we want them to be in by using an electric fence. This is a thin single strand of plastic string with a trace of wire plaited into it, supported by insulated steel posts every ten yards, with the line of string set at hip height. Either a mains or battery-operated electric fencing unit sends an electric pulse of 10,000 volts along its length every couple of seconds. If you go within half an inch of this wire, a spark of electricity will jump the gap, resulting in what can only be described as a sharp slap.

Just like me, cows hate it and keep a healthy distance from the risk. Except Frieda, that is. She could see the lush green grass, waving temptingly in the breeze on the other side of the fence, just a foot away from her nose, and the lure would make her drop to her knees and creep forward towards the new grass. When her shoulders were a whisker away from the fence, she would stretch out her rough curling tongue and tear great tufts of what, judging by the way her eyes would roll in ecstasy, she must have thought was bovine ambrosia. With a shuffling movement, she would stretch further and further under the fence, dipping her back in an inverted bovine limbo dance, getting as low as she possibly could. Once her shoulders were under she could normally get half her back through before the fence made contact. When it did, its strong electrical pulses would kick in. Far from stopping her, they drove her forwards and then she was in seventh heaven, with all the best grass at her entire disposal.

The funny thing was, she knew she was being naughty, because if I spotted her and went into the field, switched off the fencing unit and scolded her, she would come trotting over to me with an expression of complete innocence, as if to say, 'who, me? It really wasn't my fault!' And she would wait for me to lift the wire for her so she could walk under it. I am sure that having the wire lifted was her way of ensuring the fence was switched off before she ventured underneath it again.

Another of Frieda's attributes was her ability to open gates. She had mastered the art with all the gates on the dairy unit, which were a bolt type, and she was so quick at undoing them we had to wrap the old cow chains around the post to prevent her. These defeated her but did not prevent her spending many hours nosing them, licking them and manipulating them with her tongue. I'm convinced she had worked out how to undo a cow chain, and although her own inadequate dexterity meant she couldn't, she never gave up trying. Were these anomalies the result of intelligent reasoning, or had she just stumbled upon solutions by trial and error?

Her other party trick, and I'm convinced of this, was that she could count. Again, there was reward if she got it right. The milking parlour held sixteen cows, eight on each side. There were individual electronic feeders, which were controlled by a panel at the rear of the building. As each cow passed by, you could identify each by the freeze brand clearly emblazoned on her rump and see what cubicle she went into. You'd enter her number into the panel, which in turn would deliver the right amount of feed for that cow. The spout leading from the overhead cattle cake hopper to the sixth feeder had been twisted and damaged by a heifer the first time she'd entered the parlour. She had panicked, reared up and straddled the cow in front, damaging the feeder in the process. The damage hadn't stopped it delivering cake, but Frieda had worked out that if she put her tongue up the end of the spout and gave it a good wiggle, more cake would tumble down into the trough. She was bright enough to realise that wiggling it for a long time was far more efficient than stopping and starting and only eating the few nuts of cake it delivered if she gave it a short wiggle.

So how did she ensure she arrived at the sixth feeder? Cows are creatures of habit, and normally enter the milking parlour roughly in the same order. Often, a boss cow will like to be first in each side but to get to the same position, twice a day, every day of the year, so that's why I'm convinced she must have been able to count.

Ashington Farm was literally on the outskirts of Ashington, the birthplace of soccer legends Bobby and Jackie Charlton, and where the street scenes of the film *Billy Elliot* were shot. The town had formed around the pit, and most of the housing was back-to-back miners' rows, to which we delivered bottles of unpasteurised (green top) milk with a horse and cart. This was topped up from a van at strategic points on the round. Milkmen using horse and carts had ceased to exist almost 20 years earlier in Ashington, but because the farm was on the outskirts, delivery this way couldn't have been more efficient.

Sally, the chestnut mare, would walk slowly up and down the rows as John the milkman placed pints on virtually every doorstep, never having to climb in and out of a van, let alone stop and start one. The customers loved coming out and giving Sally bread or carrots, so much so that John often had to bring the excess home to be fed to her later in the day. The milk round was enormous, and as I improved the herd's milk yields and put more milk into the bulk tank, we were able to extend the number of bottles of milk we could produce.

We were one of two suppliers in the town. The other was the Co-op, and we met in the middle, only the 'middle' for us kept on moving, with a few more of our pints placed on more doorsteps every day. The good people of Ashington preferred the farm fresh green top and readily accepted the change in supplier, but the opposition weren't happy. Milk rounds had a value. The goodwill was traded at about £50 a gallon, and, naturally, dairymen hated losing their trade. In reality, it had little to do with either supplier, because it was the customers' decision, but for the Douglas brothers it was a good job that John was as fit as he was and as handy with his fists, because he often had to defend his right to sell our wares. Customers were unaware of these early morning battles on their doorsteps and within the industry. The warmongering didn't stop till supermarkets eventually destroyed the doorstep trade.

Dairy cows and duck races

One Wednesday morning I had just finished milking
and was walking out from the bulk tank room into the yard
when Sally cantered in, the cart rattling along behind her,
scattering what remained of her load. She ran up to me,
whinnying, and skidded to a halt. There was no sign of John.
Sally was agitated and sweating. I tied her up, threw a
blanket over her and gave her a feed with a nosebag. This
calmed her down and just as I was thinking I had better jump
in the van and see what had happened to John, he sprinted
into the yard, totally out of breath. As he was delivering to
his last few customers, Ernie Robinson, the milkman from the
Co-op, had walked straight up to him and had pushed him
over the kerbstone, shouting at him to leave his customers
alone, then walked off, but not before slapping Sally with
a broom handle that had been resting against a bin. Sally
took off like a rocket and didn't stop till she made it back
home, scattering her load the length and breadth of the
streets, the noise of the falling crates and the breaking bottles
spurring her on homeward, to safety. The repercussions were
momentous. Quite rightly, Ernie lost his job, and the police
also deemed that a horse and cart left to move by verbal
command was unsafe. We either had to have a two-man
team operating the round, or use a van. It was the death knell
for our horse and cart delivery, and to my mind a sad day.

My right hand man on the steading, and an unlikely friend,
was a young man (well, he was younger than me) called
Phil Davidson, and although I wouldn't exactly have called
my upbringing privileged, compared to his I was raised like a
prince. When I first met him I thought he was a young tramp.
He had a head of hair like a thick, blond, curly felt carpet,
with bits of straw and debris in amongst the pile. He was
wafer-thin and all his movements were at double speed. His
jeans and T-shirts were crumpled and if they weren't actually
dirty, they didn't look clean either. When I first met him I
thought he would be a bit smelly, but I was wrong. He smelt
of carbolic soap, of which we had a big supply in the dairy,
left over from times gone by.

Phil was estranged from his parents and had ostensibly
been in foster care from an early age, but his various foster
parents could never control him. He would run away –
literally. He would run for miles and miles and for days at

a time, only returning when he was starving. While he was on these trips he would just forage for food and drink from anywhere; bins, the countryside, the seaside etc. He only went to school when the police or inspectors took him and that was rare because nobody could catch him. And he could *parkour* long before anyone had thought to give it a name.

Phil's feral nature allowed him to sleep rough, in hedgerows in the summer and bus shelters in the winter. When I first met him he was just hanging around the farm watching what was going on. As time passed, he helped more and more as he or we saw a need for assistance, and he never asked for anything in return. He was still wild, but he had a natural affinity with the animals, and as he helped around the farm so much, Bruce Douglas offered him a job. At the start it was like working with a whippet crossed with an Irish wolfhound. You never quite knew how he would approach the jobs he was given, but whatever it was, it was always done fast. He could absorb information like a sponge, and slowly, we all realised what a bright cookie we had on our hands. Very quickly he became a valuable member of the team. He still liked to run, and at the end of an exhausting day when we were all knackered he would go for a run, and not just a little jog: it would be a minimum of ten miles.

With a bit of money in his pocket he made contact with his grandmother (where she was before, I have no idea) and he started to spend the odd night in a bed. As the months passed, he became much more personable, but he still had done nothing about his hair and he asked me how he could go about getting it cut. I of course told him, "the barbers," and he confessed he had never had his hair cut and because it was so matted, he only washed the outside. He couldn't remember the last time he had brushed or combed it, or indeed when or if it had ever been looked after. What would a barber say to him? He was ashamed of his appearance but was too nervous of people's reactions to do anything about it.

I rang up the old-fashioned barber's shop on Bell Street and explained the situation. To my surprise, the barber knew Phil by sight and told me to leave it up to him. He closed for a half-day on Wednesday lunchtime, so if I could get him there by two o'clock he would have no other customers in the shop. It took him an hour and a half to cut through the undergrowth and leave his now chestnut brown, silky soft hair in a mullet cut, even though it would be a few more years before mullets hit this mining backwater. He explained he had cut off all the dead hair and when his scalp had settled down he was to return to have a more modern style cut. He never charged Phil a penny, saying he was glad to help. The haircut transformed Phil; he was a handsome young man and it was the catalyst for him to take charge of his life and live it. A fortnight later he returned to the barber of his own volition and had his hair styled.

One day in the milking parlour, I was amused to see Phil vaulting over the partitions between the cow cubicles. These would have been four feet tall and five feet apart and there were forty of them in a row. He wasn't scissor jumping, like a high jumper or using his hands in a gate vault. He was jumping from a standstill, with both feet together, leaping with one bound from one cubicle to the next, all the way down the row and then back again. I couldn't even jump over one without putting my hand on the top bar and then I could only manage about ten. I played squash at Newbiggin Sports Centre but I didn't have a regular partner to play with, so I asked Phil if he would come along. He laughed. 'Hadaway man,' he said. 'S'ower posh, tha naas!' (Translated, that means, 'I say old chap, that's overly suave for me, don't you know.')

After a few days of me nagging him he eventually agreed to go to the last session on a Saturday night, when I assured him the courts would be deserted. I was right; we had the place to ourselves. We played in jeans, bare footed, bare chested and in all the time we played, after his first rally while he got used to a squidgy ball, I never managed to get a ball past him. From then, we played there most Saturdays. We never scored a game, although each rally was hard fought. The sports centre receptionists would just let us play on till they closed and they would often stand in the viewing gallery to watch Phil's athletic determination not to allow a ball past him, performing Daniel Keatings-style manoeuvres to make the return. We often played up to two hours straight without a break, coming off the court seeing stars, we were so exhausted, but we laughed and laughed. To my mind, that is what sport is about: camaraderie, crossing taboos, playing for the sake of the game, and not the business enterprises they call sports today. When I eventually moved on from Ashington we exchanged letters for a number of years. His were written by a girl he knew, who eventually became his wife, but as children arrived our sporadic correspondence dwindled to nothing.

And of course children bring their own rewards and challenges. The fence around our garden was constructed to be childproof. A metre into the bull paddock next to the house was a second, electrified fence, which was there to keep a bull away from the childproof fence. After we had been there for nearly two years and settled into a daily routine, Bruce Douglas travelled down to the Bath and West show, where he bought a prize-winning Red Devon bull. It was a monster, weighing at least a ton. He was to be the new sire for the beef herd on one of their other farms, but while he settled down from his long journey he would be living next door. Devons are a pure beef breed and therefore normally extremely docile. Our new neighbour was no exception.

The term "child proof fence" didn't apply when it came to Sharron, who, at the age of 22 months could bimble around confidently on the flat, but on a field that had numerous divots and large clumps of uneaten grass was decidedly unstable without a hand to hold onto. One afternoon, we were outside gardening. We turned round twice and Sharron had disappeared. We ran around like headless chickens looking for her, but she was nowhere to be found. Of course, we didn't look on the other side of the childproof fence because it was, well, childproof. As we widened our search, time was rushing on and our anxiety levels were going through the roof. I rushed outside the back gate, spinning on the spot not knowing which direction to run, when through the blur of my panic I spied something colourful directly beneath the Devon bull. I leapt over the fence to get a better look and the bull swung his enormous head round in my direction to watch me. I froze when I saw the nightmare scene; standing directly beneath the bull was Sharron. Halfway along a bull's belly is a tasselly bit where his half-metre long todger exits and Sharron was stretching both hands above her head, banging the tassel with all her might.

If I had approached the bull, he would have moved and inevitably trampled my little angel. So what to do? In a sweet and calm voice (not my normal fog horn), I said, 'Sharron, do leave the nice bull alone and come to Daddy.' Sharron was an independent soul. In fact she still is, and to be honest I didn't think she would move, but to my amazement she said, 'Daddy!' and she lifted up her arms and lurched towards me. As she moved from under the bull, he moseyed away and I swept her up into my grateful arms. To this day I feel that the bull was aware he had something small and vulnerable beneath him and had to stand still. It has always amazed me how intelligent cattle can be as long as they have the opportunity to stay calm and are given the time to work things out.

All dairy farmers gather a lot of knowledge about sex and reproduction, not only with their bovine charges, but with the couples they stumble across in their fields. How they acquire knowledge of the latter is quite simple to explain. Cows are gathered late afternoon for milking and then returned to their grazing in late evening. Similarly, we visit calving cows during the wee small hours, and we are likely to find *al fresco* courtships well under way as we pass by. Sometimes we would have to ask the occupants of cars to move out of gateways, while at other times we would have to ask couples who were lying in fields to stop, as they were at risk of being trampled by cattle, in midst of their passion. Normally this triggered an immediate response as they leapt to their feet with red faces, men pulling up their trousers and the women smoothing down their dresses. Levels of chivalry and boldness varied greatly.

These encounters always happened late evening or at night, except for one, which happened one early morning in mid-June. The night grazing field for my girls was a medieval rig and furrow that had a preservation order on it to save the deep hollows and mounds that are the feature of an ancient field like this. When I collected the cows for milking I had to cover every inch of the field, because my girls would lie hidden at the bottom of the furrows, reluctant to get out of bed, much like women all over the world. (This may sound confrontational, but I believe it to be true: you tell me of any woman who will voluntarily get out of bed to have her breasts fondled, with no possibility of it leading to other things!)

Anyway back to the story. It was a glorious dawn when I made my way to the far end of the field, calling all the way for the ladies to rise and shine. I have to admit I am not a quiet person. I have voice like the Queen Mary's foghorn, and that day I'm sure most of Ashington knew I was up and about. As I was about half-way down the field, I noticed two neatly folded piles of clothes on the top of a rig. I walked over to where they were with trepidation as I expected to find dead bodies. To my mind there didn't seem to be any other logical explanation. How wrong can you be? In the bottom of the furrow was a couple, absolutely stark naked. He had her hands pinned to the ground above her head and her legs were wrapped around his back. Judging from their expressions, they were having a whale of a time. I stood for a moment, thinking I would just leave them to it and started to make my retreat. Then, the man noticed me and without missing his thrusting rhythm said, 'good morning!' She twisted around and bid me 'good morning' as well, just as if we were passing on the street. At no point did they stop or even break tempo. But the real surprise was their age; they must have both been well over pension age, and clearly very fit, lithe ... and a couple of game old things!

7 Italian neighbours and a career opportunity

Sergio Margarelli was, and possibly still is a bottom-pinching Italian who had real charisma, swarthy good looks and the dazzling smile that beguiled both male and female alike. He owned the best fish and chip shop in Snodbury, which was so popular that from the moment he opened at lunchtime till late at night he had a queue of customers, often snaking out the door. Sergio was a man who was happy in his work. He would sing Italian opera in between his heavily accented conversations with his customers. Actually, they were more like interrogations. 'Heya Meesses Pauly,' he would say to a mother. 'How's-a your-a boy? He'll-a be a big now?' Or to an attractive lady, 'Allo bellisima, I wus a wonderin' where the sun she was and then she comes into ma shop!' He was full of it.

His fish and chips were superb and his menu was vast and inventive, but the people of Northumbrian mining towns are very conservative in their tastes and cod and chips was the order of the day, so when I started choosing plaice and chips he was thrilled. He would cook a dinner plate-sized flatty, all golden in his special batter. Then, because he was so proud of it, he would place it on greaseproof paper, come out from behind the pans and show it to the other customers standing in the queue. 'That's-a fish, eh?' he'd say. 'It's-a beautiful, no? I'm-a so gud at-a the cooking!'

The population of Snodbury lapped it all up. They never cottoned on to the fact he had never been anywhere near Italy, and when he was at home he had as strong a Geordie accent as anyone else in the town.

The Margarelli family consisted of his wife Carmella, their two children and Sergio's father, Franco. He was our neighbour, living about the best part of a mile away on top of Colridge, the highest hill in the area, in a tiny Northumbrian stone farm worker's cottage with a range of outbuildings.

The place had loads of potential for extending the cramped living accommodation. Sergio and Carmella had plans drawn up to do so, but the building work meant losing some space in the meantime, so Sergio asked his dad if he would move out while work was being done.

The plan was that Franco would live in a flat in the nearby town, which was about half a mile away at the bottom of the hill. During this time, Franco would be able to have a pint or two at The Star Inn, buy a daily paper and generally have a break away, a bit of a holiday from the isolation on top of a very steep hill. Previously his social life had revolved around The Leek and Whippet, which was once a spit and sawdust real ale pub about 100 yards away on the crest of the hill, where you could hear impromptu concerts with local musicians jamming anything from North Country ballads to AC-DC. Unfortunately its fabulous location and large car park had attracted the attention of the plastic pub brigade, who brought in a chicken-in-a-basket type menu. This had all but got rid of the locals, including Franco, who now only used the facilities in an emergency. His daily walk to the pub gradually decreased until the sight of this distinguished, slightly stooping man in his brown felt Russian type hat, swinging a silver topped black lacquer cane, all but disappeared. He became reclusive, barely venturing beyond the wooden five bar gate bordering his property, but he would often be seen leaning on it, gazing pensively up the track leading to the main road and the great wide world beyond.

He point blank refused to move to a flat in the village.

Sergio's wife, Carmella was a psychiatric nurse, small and extremely attractive, with short blonde hair, who juggled shift work with the mothering of her and Sergio's two primary school-age daughters. She also had a very short fuse and she was desperate for more space.

Sergio was the sort of character who nearly always got his own way, no matter what the circumstances, and when he didn't, his fiery Mediterranean temper would erupt like Vesuvius. Carmella got on as well as could be expected with her father in law, but life was all too cheek-by-jowl for her, and because he refused to move out, they could not proceed with their plans. The row that erupted between her, Sergio and Franco over his refusal to move was of monumental proportions, but he would not be shifted from his decision. So, Sergio and Carmella decided to buy a semi on the outskirts of Morpeth and leave the stubborn old git, as Carmella had christened Franco, to fend for himself.

Just a few short weeks before moving day, Sergio, Carmella and the two girls packed their belongings into a hire van and moved out from rural life on Colridge. Two months after their move, when Carmella had put curtains up, positioned new furniture and laid new carpets – in short, had settled her family in to their new life, they received a telephone call from Franco. He told them he was moving out from Colridge to a nice little flat in Newbiggin and that the house was now available for renovation. Of course Sergio and Carmella had used all the money they had saved for the renovation to buy their new house. They were committed to their new lifestyle and renovating Colridge was now impossible. Carmella took this news in her stride. Her life had been made simpler, she was near to the schools, her work and the shops and she didn't have to put up with Franco living under her feet, but Sergio was almost catatonic with fury. The cottage at Colridge had been his family home and he loved the place.

At this point in the story, I should lay down a Scale of Annoyance; a sort of Richter scale, going from 1 to 10, to give you some idea of how Sergio's temper grew in the following events. As I am sure you can imagine, Sergio was indeed infuriated and on the news that Franco had decided to move out of Colridge would have registered his annoyance at level 1. A week before Franco's move to Newbiggin, Sergio took a pile of empty cardboard boxes to his father, deposited them inside the front door and with a cursory growl, instructing him to pack his belongings in them before the following weekend, because, at 8 o'clock precisely on the following Saturday morning, he and a friend would arrive with a borrowed van to transport Franco's worldly possessions to his temporary home. They had to return the van before 1 o'clock, so Sergio told Franco he must be ready on time.

At 8 o'clock on Saturday Sergio, friend and van arrived at the front door to find it locked. When they opened the door, they found Franco, still in his pyjamas. The pile of empty cardboard boxes hadn't been moved, let alone filled. Sergio swallowed hard. A pulse in his temple started to twitch and his temper moved up to 2 on the Scale of Annoyance.

Using growls rather than words, Sergio sent his old man upstairs to shave and get changed into sensible clothes, while he and his friend carried boxes into the kitchen to start filling them. Throwing open the cupboard door, Sergio could not believe the sight that met his gaze, or the smell that hit his nostrils. Over the previous three months, Franco had worked his way through every piece of clean crockery, then put it all away unwashed. There was mould on it that Alexander Fleming would have found intriguing, if not downright useful. In fact he was sure there were new species that should have been registered with the Natural History Museum. Sergio would now have to wash and dry the crockery before he could pack it. That is, if he could find the sink under every available pot and pan that was piled in and around it. This pushed Sergio to Annoyance Level 3. His temple started to throb. But there was no washing up liquid. Sergio reached Level 4 and the throbbing in his temple turned to a strong, visible pulse.

Sergio asked his mate if he would start washing up, using an old bar of soap he found, while he went upstairs to see his Dad about the midden that was the kitchen. On his way upstairs it seemed prudent to carry some of the empty packing cases and while he was doing this, thoughts of his ageing father living alone for the first time in he did not know how long, entered his mind. He realised he hadn't been back to visit properly since his family had moved out and some of his rage began to wane as the thought pricked his conscience.

As Sergio reached his father's bedroom door, it opened and there stood the elderly gentleman, still in his dressing gown, the picture of abject misery, holding his arms out in front with his palms turned up. 'Son,' he said, 'I am so sorry.'

Sergio's anger completely disappeared. Thinking his Dad was an old school macho Italian whom he had deprived of female care by moving away, he put his arm round him and told him not to mind. Now that he was upstairs the pair of them would pack his clothes. Franco told him that would be nice, but he first had to go to the toilet. As Franco disappeared out the door, Sergio opened the wardrobe. He was puzzled to see that it was empty, apart from a few sports jackets. He turned round and saw a double blanket on the far side of the room covering a large dome-shaped pile. 'No, it can't be,' he thought. 'Surely not.' He pulled the blanket back to reveal all of his father's clothes, all dirty, in a pile. That was why his dad was still in his dressing gown. Sergio now reached Annoyance Level 5 and he let out a few well-chosen words that called his father's legitimacy into question.

He was going to need help to sort all this lot out. Clearly the move was not going to happen that day. All this washing had to be done before they could move, so they carried it down to the kitchen. It was then they discovered that the washing machine was broken. According to Franco it had been like that for nearly two months. Reaching Level 6, Sergio phoned Carmella. She arrived half an hour later with the girls in tow, appraised the situation and took charge. She packed the dirty laundry into plastic bin liners and dispatched Sergio with them to the Washerteria. She also told him to take his daughters, because while the laundry was in the machines he would have very little to do, so the opportunity of spending quality time with his children was too good an occasion to miss. Meanwhile, she would whip the remaining male workforce into shape and have the place shining like a new pin. By the time they had finished, her calm, Zen-filled husband would have returned with the clean, dry washing to complete his original mission of moving his father to his temporary home.

So, there was young Sergio, driving down the narrow winding hill to the Washerteria. This emporium was sited in the middle of the parade of shops (parade is a very generous description of the damp, dowdy shops that serviced this once bustling, but now quiet and desolate community, made so by the on-going mine closures). All the parking close to the Washerteria was taken, which meant Sergio had to park in the public car park, several hundred yards away. It him took three trips to fetch the six bin bags, which were splitting under the weight of the festering clothes, to the Washerteria, with his two little girls dancing around, emitting high pitched squeals of delight at the thought of visiting this wonderland. Imagine a pressure cooker, fitted with its 15lb weight when the heat is turned to maximum. Then imagine that moment just before a stream of steam screams out beneath the stainless steel weight. Sergio was that pressure cooker and he had got to Level 7.

Safely inside the empty washhouse, Sergio set about his task. There were three washing machines on one side of

the premises and three large tumble dryers on the other. At
the far end there were vending machines, one selling soap
powder and the other giving change for one-pound notes.
On two of the washing machines hung a sign saying, "out of
order." Sergio took a deep breath and placated himself with
the thought they were big machines and they would hold
two bin bags each. The rapid wash took 30 minutes, which
he thought was acceptable. Not good, but acceptable.
Just. How much was it? 85p. He had plenty of £1 notes and
a few 10p pieces, but thanks to Sod's Law, no five pence
pieces. He went to the change machine and fed it a £1 note.
It gobbled it up greedily but gave him nothing in exchange.
Under normal circumstances a person in a reasonable
temper would not risk another pound, but Sergio was a
boiling cauldron. Common sense had abandoned him and
he took the risk of feeding the pound-eating monster with
another, which again it swallowed, with a distinct clicking
sound and nothing else. So he punched it. Nothing. Then
he shook it. Nothing. So then he kicked it, and it crashed to
the floor amidst a cascade of plaster and cement. The girls
thought this was great, and bounced around, kicking bits of
brick around the floor.

Tears of pure rage and frustration came into Sergio's eyes.
He had his girls with him, so he had to set an example, but
while showing no more outwardly signs he passed to Level 8.
He took a long, shuddering deep breath and settled the girls
on the bench running down the centre of the Washerteria
and gave them their colouring books and crayons. They
were not to move, he told them, until he returned from next
door with some change. Next door was a Chinese takeaway,
which at 9.45 on a Saturday morning was not open. Next
to that was a corner shop, open till midnight but again, not
open at 9.45am. Across the lane, The Commercial public
house was closed. Next to that was a cobblers, (closed) then,
the Northern Rock Building Society, also closed. Next, wonder
of wonders was a newsagent and general store, which was
open, but on the door was a notice that read, "no change
given for the Washerteria." Sergio's heart sank. Then, he had
a brainwave. He could buy his washing powder there and
then they would be forced to give him change.

Armed with lots of change he ran back to the
Washerteria, very aware that he had left his two little angels

by themselves. All was well with them and he quickly set the machine going with the first load of washing.

Unfortunately, in his masculine ignorance he had bought soap powder for hand washing. The amount of bubbles that squeezed out of the seal of the top-loading machine was alarming, but Sergio put it down to something his father had spilt onto his clothes. Two extra rinses and another half-hour later, the first load was finished. He transferred the wet stuff into the tumble dryer and set his next batch off in the washing machine, playing I-spy and ring a ring a roses with his increasingly bored, thirsty and hungry children. Towards the end of the extra rinsing of the second batch, the door to the laundrette swung open and in walked a woman who could have been Les Dawson's double. She went over to the two machines with the "out of order" signs on them, opened their lids, took her washing out and transferred it to the two unused tumble dryers, fed her money into them and set them running. She then returned to the washing machines and removed the out of order signs. Folding them neatly, she put them in her pinnie pocket and walked back out the door. It took Sergio a few minutes to understand what had just happened. The woman had put the signs on the machines so that no one would interfere with her washing; it also almost guaranteed that the two tumble dryers would be empty and ready for her to use. She had cost him two hours of his precious time.

He was practically moaning with the anger that was boiling inside his head. He had reached Level 9, but at that moment the ping of the timer brought his mind back from murderous thoughts to folding the first dry load into a nice neat pile. He then transferred the washed clothes to the tumble dryer and started the last load. The room was becoming increasingly hot with all three tumble dryers chundling round and the girls were getting increasingly bored. 'Are we finished yet?' They said. 'When are we going home?' 'I'm really thirsty!' 'Kristy hit me, Daddy!' 'No I did not, and anyway she snapped my pencil!'

Finally, the last load was processed and Sergio was busy folding it up. The girls were quiet, having found something to occupy their inquisitive minds, at the end of the dryers in the far corner of the Washerteria. Suddenly, there was a crash and a big grey cloud of dryer fluff exploded into the room, covering everything it touched. And It touched everything – people, walls, machines, newly-done washing, clothes, and the floor. It was impossible to breathe, so all three ran outside covered in downy fluff, looking like newly-hatched emus. It took a good quarter of an hour for the dust to settle before they could go back to retrieve their belongings. This gave Sergio time to reach the giddy heights of Level 10. What he realised, when he looked around, was that his darling daughters had managed to remove the cover of the fluff filter at the back of the tumble dryers, and the blowers from the three machines had blasted what was several years' worth of fluff and dust into the atmosphere.

Washing and children having been unceremoniously shoved into the back of the car, Sergio drove back to Colridge like Paddy Hopkirk on the final section of the RAC rally. He skidded to a halt on the gravel drive behind the van to be greeted by his father, still in his dressing gown, arms spread, looking like the statue of Christ the Redeemer that stands over Rio de Janeiro. 'Son, I'm so sorry,,,' he began.

Sergio, seeing the dressing gown, was tipped beyond Level 10. The red mist had come over him. Something had to give. And it did. Sergio walked up to his father, punched him squarely on the jaw, and sent him crashing to the ground. Sergio just strode past him, leaving him to recover and struggle to his feet.

Six weeks later Franco was dead. He had been riddled with cancer and must have been in constant pain. He had never consulted a doctor. The remorse and guilt that Sergio felt was palpable.

While they were waiting for the body to be released, Sergio and Carmella went through Franco's personal papers. Sergio was keen to find out about his Dad's history. He had been an intensely private man and never discussed any of his early years with his son. Sergio knew Franco spent some time in the USA, but did not know where or why, so he was flabbergasted to discover his father had been a stunt man in Hollywood. There were professional photographs of him in various costumes, posing with actresses such as Olivia de Havilland and Anne Sheridan.

Sergio found it unbelievable that his Dad had simply sat on top of Colridge and had done nothing for as far back as he could remember. Under the photographs, wrapped in tissue paper, was the Robin Hood costume Sergio wore in one of the photos. Under the costume was a brown envelope. Written on the document inside was the reason the Margarelli family had spent their days on a hill in the middle of nowhere. It was an agreement between Franco and a Don: not a man's name, but a title. It said that if Franco spent the rest of his days on Colridge, his wife's, son's and his life would be spared and he could keep what little funds he had, but he was never to show his face outside of Northumberland. All deals would be off if he did. It read like one of his father's film scripts, except one of the two signatures at the bottom was clearly Franco's. The ink they had used was brown and smudgy. Was it blood?

The day of the funeral arrived, and it was to be a lavish affair. Sergio had announced his father's demise in the Newcastle Evening Chronicle and he had been staggered by the amount of condolence cards that had arrived; not in their tens and twenties, but hundreds, all with long Italian names and all asking for the funeral details. Sergio arranged high mass at the big Roman Catholic Church in Morpeth and then a service at the crematorium. I attended the funeral to support Sergio and Carmella. I had only known Franco for a couple of years, but he had become part of my landscape. Although he was a man of few words, the few I had passed with him had been meaningful.

Sergio was on an immense guilt trip and wanted to give his father the send-off he deserved. The church was full of flowers, the cortège was ostentatious. There were even attractive professional female mourners who draped themselves over the coffin and wept as we waited for the service to start (after the funeral, Sergio denied paying anything for them and had no idea who they were). The church was filled predominantly with short, stocky, balding men in black suits, all wearing sunglasses. All they needed were the violin cases to complete the picture.

After the hour and a half of Latin and incense, the pallbearers arrived to move the coffin back to the hearse for transportation to the crematorium. The cortège left the church at 11.30 to get to the crematorium, which was out of town. They had to be there by 11.45. Wednesday morning in Morpeth is a market day, and there was also a miners' picnic (a bit like a country fair) going on. It was busy to say the least. Everyone had to make their way through the one-way system and through the traffic lights. It turned into wacky races. All that was needed was the Benny Hill chase music. The signposts were unclear, and a lot of the mourners – there were hundreds of them from out of town, were annoyed that no one was paying them the respect they deserved.

As I was going through the door of the chapel at the crematorium a man in a broad fedora turned and said to me with a thick Italian accent, 'My word, but it's hot in here.'

Putting both feet into the mire, I said, 'what do you expect? It is a crematorium.' Realising what I had said, together with the solemnity of the occasion. I started to giggle. When I start giggling, I cannot stop, and to make matters worse I am a very noisy giggler. The vast number of people trying to get into the chapel prevented me from escaping until I had got my infantile affliction under control, so I took my hanky (luckily a nice clean white one), out of my pocket and stuffed it unceremoniously into my mouth.

I found a corner at the back to sit in. I was totally out of control. Tears of laughter were running down my cheeks, I was making hissing noises and I was trembling. I hung my head to deaden the sound. Mercifully the service was very short but to my dismay I saw Sergio making his way to me. I took my hanky out of my mouth and clutched it in my hand, clenching my fists till the nails dug into my palms and pulled my face into a more contrite expression, keeping the urge to start giggling again strictly under control. Sergio said, 'Philip, I'm touched by how much you must have thought of my father. Thank you for coming. I'm sure he is in a more comfortable place.'

He had clearly mistaken my appalling behaviour for somebody who was overcome by grief. To this day I have never told him the truth, mainly because I still get the giggles if I think about it.

When we had been at Ashington Farm for a couple of years, I saw a full-page advert in *Farmers Weekly* for a Dairy Farm Manager in Saudi Arabia. The salary was lottery money. The dairy units, newly set up in the desert, were unique and untested. The advert said that until all the facilities were completed, life for the staff at the facility was to be very basic.

Masstock Systems of Northern Ireland, a large international company, had secured the contract for setting it all up. Without hesitation I sent for the application form. The salary they were offering would give me the opportunity to earn enough money in a few years to set us up in our own farm. In addition, the excitement of doing something adventurous was incredibly appealing. Married accommodation was available, and if we stayed there for three years we could return to the UK in time for Sharron to start her education. A few days, later a parcel arrived for me. I tore off the wrapping paper to reveal an A4 folder containing about 100 pages of questions. I know now it was a cleverly designed psychometric application form, but at twenty-two years of age I just thought it was weird. Consequently I filled it in the way you are supposed to, writing down the first thing that came to mind as answers. It took me about an hour from start to finish. I have terrible handwriting, and my spelling was, and still is atrocious. Once I'd completed it, I parcelled it up and sent it back to Belfast, not expecting any further contact.

A week later, to my utter astonishment I received a phone call from a lady with a very sweet Irish accent, who asked me if I could come to Belfast for an interview on the following Friday. I was dumbstruck for a few moments and then I gathered my thoughts and said, 'yes. I will have to find ferry times, but I am sure this would be possible.'

To this, she said, 'there's no need to be bothering with all that. Sure, the tickets will be waiting for you at Newcastle airport and we'll be doing everything else. I'll be letting you know times and such like in the days that follow.'

At 4 o'clock the following Friday afternoon I flew from Newcastle to Belfast airport, where I was met by a chauffeur and taken to a five-star hotel. I thought to myself, 'this can't be right.' I'm a small time herdsman with very new qualifications and not a lot of experience. When I signed in, I was asked to meet the other candidates for dinner at 7.30, but until then there was an information pack in my room to read and familiarise myself with, prior to my interview the following day. I was majorly outside my comfort zone, but I thought if my young friend Phil Davidson could take the strides he had, then I should "man up" and play the part.

That evening, the dining room was buzzing with the conversation of a large group of agricultural intellectuals, all shiny brown brogues and loud guffaws, but as I entered, prompt at 7.30, a sudden stillness descended on the place. A Colonel Blimp-style character separated himself from the group and said to me, 'I'm afraid young sir, this is a private dinner party for Masstock Systems.'

I must have been a clear twenty years younger than anyone else in the room. 'Yes, I know,' I said. 'I'm here for the interviews as well.' This little exchange did nothing to boost my conviction that I should be here.

The Colonel instructed us all to take our seats and asked us all to stand one by one and introduce ourselves, giving a brief history of our background. They started recounting how many hundreds of men were in their charge, how many thousands of cows they managed and how many thousands of acres they farmed. It eventually got to me. What to do? Tell the truth like me mam always told me to do. So I girded my loins, stood up and in the broadest Geordie accent I could muster did exactly that. 'I,' I announced, 'milk a hundred and twenty-five cows for the Douglas brothers in Ashington.' Initially there was silence then one by one they started to laugh, realising how ridiculously pompous they had sounded. They thought I was teasing them. I didn't enlighten them, but I felt completely out of my depth.

The next morning, a series of limousines chauffeured us to Masstock's head office. Once there, we were split into two groups. Worryingly, I formed one group and everyone else was in another! I was taken to Alistair McGuchion, a director of Masstock who, with his brother, was the founder of Masstock Systems. He explained that there had been hundreds of applicants and they had employed some temporary staff to sort through the mail. At this my heart sank and I thought here comes the 'I'm sorry for wasting your time' part. It didn't. He carried on, explaining that he had been passing through the sorting office and he had picked up my application. On reading it, he had instructed the team that he wanted to do the initial interview with me and that I was just the sort of person to assist the farm manager. We got on like a house on fire and after two hours he handed me over to the normal recruitment team. After another day of medicals, fitness tests and practical aptitude tests they offered me a job as assistant manager, which I readily accepted.

Back at Ashington Farm I knew there was an enormous

amount to organise, such as sourcing hot climate workwear. There was a period of four months before the start date, so there was enough time to do this and study the information and paperwork that regularly tumbled through the letter box. After two and a half months I was well on the way to being ready when I got a telephone call from Northern Ireland. I was to attend a briefing in the Berkeley Square Hotel in London. As before, all the arrangements had been made, so I had to just turn up at Newcastle airport. I was met at Gatwick and given a close-typed letter explaining the reason for this meeting. The original project had been for one farm for King Faisal, but this had now escalated and all his Princes wanted one as well. Although Masstock Systems was enormous, they needed government funding for an operation of this scale. All recruited staff members were to be re-interviewed by a team selected by the Ministry of Agriculture.

The hotel conference room I was shown into was laid out like a courtroom in the Old Bailey. The interview panel of ten sat in a row on a raised area behind a large, highly polished table. I sat facing them on a small, hard backed wooden chair. They didn't actually have spotlights shining in my eyes, but they were sitting with their backs to a wall of large sash windows, putting their faces in shadow and mine in full sunshine. They told me to sit, much like Barbara Woodhouse to a disobedient Jack Russell, and the chairman asked me, with no introduction or preamble, 'what qualities would one so young bring to this project?'

Gathering my thoughts and answering too quickly in an effort to fill the ensuing silence I said, 'enthusiasm and ...'

Before I could say another word, a shrivelled, bespectacled know-it-all type asked, 'what is the difference between a lethargic professional and an enthusiastic amateur?'

I thought for a while – probably too long, then said, 'neither of your examples are ideal, but an enthusiastic professional often makes a good leader.'

The grilling continued in the same vein for half an hour. I was totally out of my depth. I knew I was not performing well, so, as one of the Ministry men (or it may have been the physiologist or the Professor of Agronomy) started to ask me another question, I stood up and said, 'Gentlemen, you and I are from different worlds and we do not speak the same language. If, for this project, you need people that know how to look after and care for cows, then I will say categorically that I am the finest stockman you will interview today. Out in the desert, in that extreme environment you will need a "stockman's eye" second to none. If you are not aware of what a "stockman's eye" is, then I suggest you find out. At no point has anybody asked about the biggest problems, which are, how are cows going to react to the heat, the inevitable increase in the concentration of salt in their diet and what levels of mastitis will result from the sand litter you plan to use at these temperatures. The first unit will have to solve all these problems and I am the man for this role. As far as I am concerned this interview is over.' And I walked out.

I never heard another thing from Masstock – no letter of rejection, no phone call, nothing. Not surprising really.

Luckily, I had not handed my notice in at Ashington Farm, so I was able to continue there. Sometime later, I heard through the agricultural grapevine that the team that was selected was made up from the academic intelligentsia of the industry. The man that took on my role of Assistant Farm Manager on the first unit was the vice-principal of Houghall Agricultural College. He was electrocuted in the milking parlour ten days after his arrival in the desert. There but for the grace of God – or perhaps not – it could have been me. As I've said before, cows are incredibly sensitive to electricity. Their bodily systems are destroyed by milliamp currents, whereas healthy humans can stand a couple of amps. Would I have noticed an electrical fault? Probably not, but the cows would have told me.

After the disappointment of not going to Saudi and the three years as a herdsman I started to look around for the next challenge, which turned out to be in the Deep South. Or rather, it was south of Scotch Corner, which for a Geordie is what Watford Gap is for a Cockney.

The disappointment of missing out on working in Saudi unsettled us, and although employment with the Douglas Brothers paid well, it was no longer challenging. Because Carol was now a full-time housewife and wasn't planning to look for outside work, we could look far and wide for the next step up the ladder, so scouring the classified ads in *Farmers Weekly* every Thursday for the next career move became our entertainment.

Then we found it.

Manor Farm

Manor Farm was a 400-acre dairy farm just outside the village of Meltham in West Yorkshire, in a tiny hamlet called Wilshaw. It had been owned by the tractor division of David Brown and it was the home of the Wilshaw Herd of British Friesians. In 1977 they sold the farm, lock stock and barrel, to the Coles Brothers, who were wholesalers to the pub trade. I got the job to manage their first agricultural adventure after a bizarre interview where I almost interviewed myself (this was becoming a habit with farm interviews). Canny businessmen they may have been, but they were entirely ignorant of farming and agricultural practices and didn't have a clue what they were doing. After a relatively short interview they offered me the job, and I found myself associating with southerners in the land of the classic BBC TV comedy series, *Last of the Summer Wine*. (It has always been a delight to me to call Yorkshiremen southerners. This is not discriminatory against the south; it's that Yorkshiremen hate being called southerners.)

I was employed as a working farm manager and Carol would often run the office or drive a tractor to help out. Just before we moved to Yorkshire she had given up her job in the accounting firm in Newcastle and we had decided that she would run the accounts side of the business to give us more together-times and even perhaps have a few more babies.

I took charge in the summer of 1978, just as the trades unions' strike actions against government wage restraint were taking effect. All the transport industries were going on strike and picket lines blocked the feed mills for my cattle feed. I had about 700 head of stock to nourish and no way of accessing anything in the quantities necessary. Kellogg's had a Corn Flakes factory in Huddersfield and I rang them up to see what they did with their waste. I was put in touch with Howarth Winterbrook, a firm of food brokers that dealt with food by-products. They were delighted with my call, as all sorts of faulty food products which had ingredients missing were being put their way because the non-stop factory processes had been upset by various strikes. As a consequence, large quantities of reject products were ending up in skips and thirty-ton tipping wagons. For some reason, the men on the union picket lines didn't interfere with these waste wagons and I ended up with human grade fodder for my cows. It was always a challenge, because I didn't always know what we had been sent until the lorry tipped the load out on our yard, and often the deal was struck after I'd seen what I'd got. Very rarely did anything end up on the midden.

All sorts arrived, like 25 tonnes of wet, uncooked cornflakes, skip loads of warm bread or broken biscuits. One day, in a 30-ton tipper, there was the biggest Weetabix you have ever seen. It slid out of the trailer on to our yard. It measured 30 feet long, 9 feet wide and 8 feet high and was made up of thousands of little Weetabix all stuck together. This giant breakfast treat weighed 2 tonnes and was incredibly difficult to break up. In the end we broke chunks off with large pinch bars, piled them into a muck spreader and blathered them into bite size pieces against a concrete wall.

You may wonder if it is safe to feed cattle such weird and wonderful foodstuffs, and certainly many of my conservative fellow farmers thought I was foolhardy. They were only happy if the concentrated part of the cows ration arrived in a recognisable format. On the other hand, I was happy with the fact that all my feeds were made of almost exactly the same material, even though they arrived wet or partly cooked. The skill was in balancing all the different elements to make a nutritious and healthy diet. Cows are ruminants, and this means that before their four true stomachs can digest their food, all of their food has to pass through a large fermenting organ called the rumen. This is filled with millions of bacteria that break down the nutrition into simpler forms so the other stomachs can digest it. As long as you introduce new feeds slowly, allowing the bacteria in the rumen to multiply up in sufficient numbers for them to digest it, you can feed cows almost anything, provided it is wholesome and vegetarian.

One day, a large flatbed articulated wagon arrived at the

farm, laden with brown paper sacks filled with unwrapped Cadbury's Creme Eggs. With all the fat and sugar in them, they were an incredibly concentrated food and it took us three weeks before we could get all the cows to crunch their way into a whole one and not just lick them around the troughs. Eventually, they liked them so much that it got to the stage where all the staff would have a pocket full of eggs to hand feed to their favourite animals. I fed 60 tonnes of the eggs to our cattle. I tried to get in touch with Cadbury's advertising department because I thought it would make a unique advert, but I never got any response.

We stored the eggs in a barn near the house and if anybody forgot to shut the door, it was an open invitation to Tosca, my finger-eating spaniel. He would rush straight in, fill his belly to bursting, drag himself outside, throw up, and then go back in for more. He never did learn self-control where food was involved.

Dairy farms often have milk that is not suitable to sell. The first milk from newly calved cows, called colostrum, is designed to give the new-born calf a nutritional kick-start. It is concentrated and contains a laxative and large quantities of antibodies. The modern high yielding cow produces far too much of this and, if allowed unlimited quantities, the calf would make itself ill by gorging on far too much of it, so the excess quantity of colostrum is put to one side. Also, blood can contaminate milk. For many reasons, a cow's teats can become damaged and these injuries can contaminate the milk with blood, which makes it unfit for human consumption. So, rather than throw either types of milk down the drain we often kept a pig or two to drink it. Milk fed pork is delicious.

Often, we kept the pigs for too long and we ended up with chops the size of brontosaurus steaks, but the crackling that came on the meat really crackled. One pig that was given to us, we called Snuffles (it's always a mistake giving a name to an animal you intend to eat). She was the runt of her litter and while I was buying two weaners as house pigs from the neighbouring pig farmer, he put her into the palm of my hand and told me that if I could keep her alive I could have her. We put her on a small towel in a shoebox on top of a warm hot water bottle in the kitchen next to the Aga, where she was as snug as a bug in a rug.

Getting her to eat was a real problem, and, for about three weeks, she was force-fed with pipettes full of milk. This kept her alive, but only just, as she seemed to be getting smaller by the day. Exasperated, I sat in the kitchen looking down at this little scrap wondering what I could try next, when absent-mindedly I picked up an apple from the fruit bowl and started crunching my way through it. Snuffles' little nose started twitching, and, with an enormous, effort she lifted up her head. I took a bit of coarsely chewed apple out of my mouth and offered it to her, and to my delight she started eating it. For the next few days I gave Snuffles small portions of whatever I ate, chewing it first. After three weeks she was eating all the scraps from our meals. A litter tray was put into the corner of the kitchen and after two misses, she used it perfectly. She was cleaner than any puppy or kitten I have house trained.

The bizarre thing was, in six months she never grew any bigger than a pug, and so became a house pet. In the evenings she would join us in the sitting room while we were watching the TV. Our two house cats would be stretched out next to one another on the rug in front of the fire, luxuriating in the heat, when Snuffles would trot in, her little feet tippy-tapping over the wooden floor. She would go straight over to the cats and lay herself across them. Although Snuffles was small she was very solid, and she was impervious to hissing cats and their claws. By the time the cats had peeled themselves from under this porcine intruder, Snuffles would be fast asleep, snoring gently.

A few months on, for no apparent reason, she started to grow, and over the following few months she grew beyond house pig size. But before she was moved from the house to the barn, she had become a useful member of the household. We were hosting a stalwart Yorkshire couple, Bert and Ethel Posselthwaite, who had been Good Samaritans to Carol after her car had broken down in the middle of nowhere and they had given her and Sharron a lift home. They invited themselves to dinner the following week and, although they had been very kind, they had no idea of life on the farm and the hours that we worked. They thought six o'clock in the morning was still night time, whereas for me, being in bed at that time was a bit of a lie in. Also, we had absolutely nothing in common with them; Bert was an accountant with the Yorkshire Water Board and was henpecked to death by Ethel.

That evening, we had made our way from the dining room to the deeply cushioned chairs in the sitting room. By ten o'clock my eyelids felt like they had lead weights on them, but Ethel was droning on about what colour her curtains were and how she had bought them from the high-class haberdashery on Huddersfield High Street. 'You know the one dear,' she said. 'It has all those exclusive fabrics in the window.' I could tell by the expression of exasperation on Carol's face that if we were to survive this we would need some caffeine, so I got up from my comfy chair and offered them coffee. As I opened the kitchen door I was greeted by Snuffles, who was eager to meet our new friends. I thought, 'why not?'

I had pushed the kitchen door so that it was almost closed, but not clicked shut. Our not-so-little piggy tippy-tapped across to the door, hit it half way along with her hard snout and it slammed open against the wall with a bang. Seeing the couple lounging on the settee she bounded over (I'm not entirely sure how a young pig bounds, but she managed it), and launched herself onto the cushions beside them. She then used her piggy rooting skills to thrust her extremely tough snout between the Posselthwaites' backsides and into the gap where all your loose change disappears, and charged along its length under the cushions. This resulted in our erstwhile Good Samaritans being ejected off the sofa and onto their knees on the floor. Picking themselves off the floor, straightening their clothes, and ignoring the tray of coffee I had brought in, they said, almost in unison, 'We really must be going!' and, 'no, of course it isn't the pig. It's such an unusual pet.'

I got a result, but boy, did I get a bollocking from Carol.

Dairy farms seem to be a magnet for animal waifs and strays, be it through a passing motorist handing a half dead pheasant over to you, some other sort of creature, wild or domesticated, dragging itself into the yard, or you finding a poor starving scrap of a creature, curled up in warm, dry straw in the barn. Sometimes they arrive as bold as brass. The morning after the Posselthwaites had beat their hasty departure I heard a scratching at our front door and when I investigated, I saw the biggest moth-eaten bruiser of a ginger tom sitting on the doorstep, yowling at me. Sharron asked me who was at the door and I said a big moggy, and that's what she named him. He was hungry and we were short of farm cats (working cats who live in and around the farm buildings), so I led him round to the farmyard and gave him a bowl of milk (a saucer wouldn't have satisfied this beast). From then on, each evening, when I fed our other three farm cats, he was first in the queue. It is a fallacy that a hungry cat catches more mice and rats. You have to feed them enough so they aren't hungry, but not enough to send them to sleep. The aim is to have them out and about looking for game to play with. That way they don't gorge themselves on their first victim, but will torment several, unfortunately, during the course of a night's work.

Moggy was a character and a half. He would crouch on the top of the stone gatepost of our garden gate with his eyes closed, pretending to be asleep. The road in front leading to the public footpath was a favourite place for the doggy people of the village to walk their pooches. Following the country code, they would have their charge on a lead as they passed the farm entrance. Moggy would wait for them to pass, then silently slide off his perch and stalk up behind them, unsuspected. When the moment was right, he would leap onto the root end of the dog's tail and sink his teeth in it as deep as he could, then sprint off in front of them. Invariably the dog owner would unleash their hound, because no cat was going to humiliate their beloved, but unfortunately they had fallen into Moggy's trap. The dog would streak after Moggy, who would keep ahead until they reached the gravelly bit of the road. Then, this ginger monster would stop with his back to a ten-foot high stone wall, turn towards the oncoming canine freight train, arch his back, raise his hackles up and positively snarl. You'll know the scene in the *Tom and Jerry* cartoons, where Tom chases Jerry around the corner and finds the little mouse standing beside a snarling bulldog. Tom tries to skid to a halt, but collides with the dog and they wind up in a Whirling Dervish of a fight that would see Tom come off worse. With Moggy, the result would be the opposite. He'd sink his claws deep into the cheeks of his pursuer, then jump clear and scale the wall leaving behind a dog so furious he would often be spinning on the spot, barking furiously.

Moggy had another trick up his sleeve, which involved one of the stone staircases that ran up the side of a granary.

He would crouch on the sixth step, which is about head height and just as you were passing he would strike at your face with his claws fully extended. All our staff had scars from these encounters.

One day, after about eighteen months of mutual distrust, he just disappeared. Where he went, nobody knows. Perhaps he returned from whence he came, or went back to the Devil. Maybe one of the dogs he tormented finally caught him or perhaps one of their owners really got fed up with this Wilshaw tiger and did away with him. One thing is for certain, and that is that none of the staff, my family or the rodent population ever missed his presence.

As a farmer you have to master all aspects of your business and gaining acumen is as important as learning the farming skills, but spotting a deal comes with experience rather than out of a textbook. A short while after the transport strikes stopped, I received a phone call from Howarth Winterbrook, the waste feed broker, asking me if I was interested in some soya meal. As a source of protein for cows, soya pretty much comes at the top of the list. 'How much is it?' I asked.

'£10 a ton,' came the reply.

Alarm bells started to ring. That was too cheap. Soya was about £170 a ton. 'What's wrong with it?' I said.

'It's red.'

'Why?'

'When it was being shipped from Brazil,' my contact explained, 'there were barrels of cochineal on deck alongside it. During a tropical storm a lot of the cochineal barrels leaked into the soya, and so was rejected by the importer.'

I thought for a moment and then asked, 'Is it all red?'

'No,' came the reply. 'Just the top six inches.'

'Where is it now?' I asked.

'On the boat at Tilbury,' he said. 'But because it's contaminated, they can't use the normal machinery to off-load it. You would be responsible for emptying the holds by whatever means you can think of, without damaging the ship.'

I needed to think quickly. I had a shed that would hold about 300 tons, which would have been a good year's supply. If I spent £60 a ton extracting it, I would still be saving £30,000. This was looking good.

'How many tonnes are there?' I asked.

'60,000 tonnes. And you've got five days to remove it.'

This represented enough soya for 200 years for my cows, and it would take 1500, 40-tonne wagonloads to move it.

I must have gone quiet, because my caller asked me if I was still on the line.

'Yes,' I said, 'but I'm sorry, but that is just too big for me. But thanks for thinking of me.'

As I was hanging up, I had a lightning bolt of an idea. I could sell this on to Rank's or BOCM and charge them £20 a ton, taking a cut of just £10. I just needed to make a few phone calls. I snatched the receiver back to my ear but we were disconnected. I re-dialled but the number was engaged. I re-dialled again and again and finally, after about ten minutes, I got reconnected to my caller, only for him to tell me he had sold it to Rank's agricultural feed division, for £30 a ton. I had missed my first and only opportunity to make one million, two hundred thousand pounds. Lesson learned: never say yes or no without giving yourself plenty of time to think of all the ramifications.

I have talked about stockmanship and a stockman's eye, and in Northumberland the Gray family were famous for their outstanding ability in these respects. So, when one day Alan Gray rang me and asked me to progeny test a young bull he was managing, I readily agreed. This meant using him on my young cows and recording how his calves performed as cows when they joined the herd. A fortnight later a posh cattle wagon arrived and a well-groomed, agile, glossy example of bovine excellence walked calmly down the ramp and into our yard. Rather than the Friesian I'd been expecting, he was a Canadian Holstein, long legged and fine boned like an Arab stallion. His progeny would be tall and give 'blue' milk, which has low levels of fat. At this time, the population of Britain was still drinking full fat milk with their morning corn flakes, but although blue milk would be worth less per litre, the progeny's yield would be double that of their British Friesian sisters.

Alan had accompanied his charge and he gave me strict instructions to work him hard, and not to mollycoddle him. He had run with some older cows that had helped him on his way, so he was not a virgin. There never has been any point in putting two *ignorami* together; they waste too much time and energy. I was to have him for six months and in that time he would have 60 of my heifers to serve and get pregnant. Also after he had finished that task – about nine weeks' work – I could use him as a sweeper bull in the herd, which meant I could use him to serve all the cows that AI (artificial insemination) had failed to impregnate.

To the uninitiated male, the bull's first task may seem the stuff dreams are made of, but having said that I am not sure even the likes of Mr Christian Grey of *Fifty Shades* fame (and being fictitious, no relation to the Gray family who were testing this Holstein bull!) would have had the stamina of servicing sixty virgins, all on a three week menstrual cycle, with heats lasting up to three days. He could be performing non-stop. Of course as each fell pregnant his task would lessen, but at the beginning, when the moon is synchronising the heifers' cycling, he would not have time to eat, drink or sleep. And he almost didn't. After his nine weeks' work, his coat had lost its bloom, the white patches in his markings were stained yellow and he moved more like Harold Steptoe than Nijinsky. We rested him for a fortnight and fed him up with a high-energy diet before we put him with the main herd.

Each cow has to have a calf every year to maintain her milk supply. To keep the genetic pool in the herd as large as possible we used the AI service, which provided semen from all over the world. At the ninety-day point after calving we would start inseminating, having monitored each cow's cycles prior to that point. Once we had paid for the first insemination, we got another two chances free to get a cow pregnant. In a well-managed herd the best you can hope to achieve is a 70 per cent success rate, but in reality it will be nearer 60 per cent, and this is when the sweeper bull is used, to try to clear up where AI has failed. For a young, lightweight bull a dairy herd is a very intimidating place. We have all heard the expression, "bossy cow." Well they all are, and unless they require the bull's services they treat him with utter contempt.

We introduced him to the herd, and to his credit he stayed the distance, living alongside his 300-strong, premenstrual, frustrated harem, all with over-large, tender udders. After his six months with us, in contrast with his original Red Rum style appearance, he looked more like one of those starving cattle you see in drought-ridden Africa. His eyes were sunken, his coat was stark, he hung his head and he was hat-rack skinny. When Alan came to collect him, I was a very nervous that perhaps I had overdone it, but all he said was, 'well, you did as I said and not mollycoddled him. He's obviously been working hard. Thank you.'

I breathed a sigh of relief.

It was about two and half years later that I heard from Alan. He told me that all the results from the young bull had been fed into the system and they were so good, the Milk Marketing Board bought the bull for, wait for it – three hundred and fifty thousand pounds!

If I had known he was so valuable I think I would have been tempted to wrap him in cotton wool, and of course that is why Alan had never told me about his provenance. I must have suspected though, because I used him on my best cow and she produced a bull, which I kept. We called him Swinney Knoll, after the only hill of any consequence on the farm, and by the time he was mature enough to work he was a real looker. One morning, when he was still an adolescent, I had forgotten to phone the AI service. You have to phone them before 10am to book a visit from them and it wasn't until the middle of the afternoon that I realised my mistake. I was really irritated with myself because, one particular cow, Jezebel was an excellent milker, but she had been having silent heats, which none of us had spotted. When a cow is on heat she gets very noisy and literally gets hotter, actually steaming in cold weather and her white hide will often have a yellow hue to it. As well as being noisy, she will get restless and to help a busy bull spot her she will mount other cows, who will run away from underneath her. Then other cows will mount her but she will stand, often having her tail head rubbed so violently, with the weight of a heavy cow pounding on her back that it would bleed. During a silent heat, none of these signs are visible.

As cows go, she was not particularly tall and although she had gone off the boil, I thought it was an opportunity for Swinney to have his first proper sexual encounter. As I said earlier, for a first-time mating it's always good if one or the other knows what is supposed to happen. Jezebel was solid in stature and when we turned her into the bull pen, she went in calmly and stood while Swinney did the sniffing and lip curling that all bulls do as the female pheromones do their work. We were a bit disappointed to see the difference in height between the two of them, and from a logistical point of view we thought he wasn't going to manage. We were then further disappointed to see him disappear into his covered area at the back of his exercise yard, apparently not interested. There can be a few reasons for this, including the cow not being properly on heat, the bull having a bad back, he's gay (yes, gays exist within the rest of the animal kingdom – humans do not have exclusivity) or more simply he's asexual.

We were just about to go in and retrieve Jezebel when Swinney sprinted out of his shelter with his hooves scrabbling and slipping on the concrete, trying to get a grip. By the time he approached his target, he'd gained enough momentum to take a flying leap and land on her back. He curled his back legs between her back legs, and giving an almighty thrust he penetrated Jezebel like a darts professional, scoring a bull's eye. This left him on her back, with all four legs off the ground and his back legs sticking underneath her udders. He could not maintain this position for long and he slowly slid off Jezebel's back and on to his, with his legs in the air. She shuffled backwards a short way and looked down onto Swinney's backside with a look of utter confusion on her face. For months, even years after, the day's work would be held up by one of us on the farm remembering Jezebel's expression, which started us off into uncontrollable laughter.

The good news was that as long as he was fertile, we knew we had a winner on our hands. As Swinney Knoll grew older, his sexual appetite grew more and more voracious. If he could get a cow to stand for him, or if he could jam her in a corner, he would mount her. As a sweeper bull he was the best. For calving, the Wilshaw herd was split into two: one a spring calving troop and the other autumn. To achieve that, we synchronise the cows' heat cycles with the use of hormones, the idea being that the AI turned up for one day to inseminate a lot of cows at once, which spread the visit cost over a large number of inseminations, thereby making each insemination cheaper. It had the added advantage of making the subsequent management of the group easier, with them all being at the same stage of pregnancy at the same time.

When a cow is on heat, apart from making a lot of noise and mounting other cows, she will produce copious quantities of glare (a naturally produced sort of KY gel) which oozes from her vagina, together with a hormone smell to attract the bull, which he can detect from over a mile away. Imagine if you will, this dramatic effect multiplied by sixty, when the whole of the autumn group is on heat at the same time, and a bull who is a sex maniac. To prevent Swinney running amok, we had to lock him up in his bullpen, which was built like a Second World War German gun emplacement. Short of a nuclear explosion, nothing was going to destroy it. He was not a happy chappie. After we had finished the evening milking we went in for tea, leaving Swinney bellowing his displeasure. At about 10 o'clock I went out to do my final look at the herd before bedtime and found Swinney had got into the bulling group. I apologise for the next bit, as I can think of no other way to describe it, but he had shagged himself to a standstill. There were forty odd cows with their tails up and Swinney was collapsed in a hot sweaty heap in one corner, fast asleep. He slept for eighteen hours and limped for a fortnight.

We found out later how he had got out of his pen. He had run flat out at a corner and then had scrabbled his way up and over an eight-foot wall, a bit like a four-legged Jacky Chan. After that, we had to put steel bars across the corners of his bullpen to keep him in.

Although it was early days for Swinney in terms of progeny testing, I decided it would be a good idea if we kept some of his semen in storage, because if he turned out to be a winner I would kick myself for not having built up a stock. Nowadays he would have been sent to an AI centre, where they have all the modern facilities for harvesting this interesting bounty. There, they use a mechanical cow, fitted with the necessary orifice. The operatives smear it in cow heat hormones and drive it around a yard, in which there is a bull that has not seen the real opposite sex for several months. The driver, who is inside the "cow," tries to get the bull interested, and eventually the bull mounts and inseminates into a warm rubber bovine-style vagina. Then the semen is divided up into tiny inseminating straws and stored in liquid nitrogen. (Incidentally, I've always wanted to know what on earth the driver of the mechanical cow puts on his census form as his occupation.)

That is the modern, safe way of collecting semen. In 1979 it was much more of a hit and miss process. Basically, I used a heated, over-lubricated, extra-large, very thick, condom-style contraption, made of rubber that's as thick as a Marigold glove, stuck on the end of a bamboo pole. Some imagination is called for here. Picture in your mind's eye if you will, in one pen, a steaming Friesian cow who is so up for sex she is bleating, and then in the other, 750 kilos of male testosterone, who would shag a wart hog if he got the chance. The idea is you put the two together and then you slip into the pen with your bamboo pole, and as the bull mounts his beloved you slip the condom over his metre-long penis. The art is that he should neither see you nor feel the device being put into position. You achieve this by moving up behind him as he mounts, but before he thrusts. You can imagine the consequences if he sees or feels you before he achieves his goal. Even, if by some miracle, you have succeeded, there you are in the middle of a pen with a very aroused bull with his penis on the end of your bamboo pole. Time is frozen, as his enormous head and neck tower above you, elevated by the height of the cow. His unfeeling coal-black eye, the size of an over-ripe Victoria plum, swivels round and he looks down to see the inside of your trembling soul. All of a sudden all hell breaks loose. As he dismounts, you slip the condom off his member and leg it for the gate. You dive through, making sure your precious cargo doesn't spill, and your assistant slams the steel gate behind you just in time for the bull to crash into it.

We managed to harvest Swinney three more times before he got wise to the situation and he would watch us all like a hawk. If he saw any sign of our intentions whatsoever, he would bellow by the entrance gate until we were far enough away for him to be sure we would not to interfere. It was probably just as well, because I am unsure how many more times I would have had the bottle to do it.

As bulls age, they develop a large fat deposit on the top of their shoulders that stretches forward to their heads. This forces them to hold their heads down. Whether or not this accompanies their newly found bad temper or is the cause of it is unclear, but it certainly seems to have spawned the expression "taking the hump." Unfortunately, Swinney didn't live to a ripe old age. He got his hump quickly, became too aggressive, and probably was the main constituent of several hundred mince pies, but before he left us, he brought another common saying to life.

In one of our small barns, the Coles Brothers kept all the glasses and crockery for their pub trade. The boxes were stacked to the ceiling in rows, with alleyways between them, just wide enough for a hand trolley. Swinney was en route to his day field when he spied the door of this barn. It was wide open. He had never been inside, so he made straight for it. Inside the barn, the delivery driver was pushing a trolley full of boxes towards the door, and met Swinney face to face on the threshold. Swinney just put his head down and kept on going. The driver, who was no countryman, had never been close to a cow, let alone a none-too friendly bull. He screamed in fright, Swinney bellowed and all hell broke loose. The terrified man managed to sidestep the oncoming bull and slipped out the door, slamming it behind him. Swinney pushed his way down the narrow alleyways, sending stacks of boxes filled with china and glass smashing to the floor. The falling boxes and crashing of breaking glass turned his curiosity into terror and blind panic and sent him rampaging around the storeroom. It took us 20 minutes to extract him, and in that time he virtually destroyed the contents, but escaped without so much as a scratch. Surprisingly the insurance company accepted the explanation on the claim form; "A bull in a china shop" without question, and just wrote the stock off. We had the task of sorting the debris out and clearing all the cullet (broken glass). We filled three skips.

Pendragon
Round Table
and more babies

Running a large dairy farm efficiently requires a good rapport with a large animal veterinary practice and a vet who is prepared to visit 24 hours a day, 365 days of the year, including Christmas day and bank holidays, and for any emergency. It's a level of commitment that the modern day GP system could do with taking on board. The vet would have scheduled visits perhaps every fortnight for routine herd maintenance, which would cover pregnancy diagnosis (PD, as we call it), fertility problems, lameness etc. If there were large numbers of PDs to perform, two vets would attend, because when you have shoved your arm up thirty backsides to feel through the colon wall for a foetus in the womb, your arm goes numb and all sensitivity is squeezed out from your fingertips.

Cameron, our vet was on the farm, with a newly qualified greenhorn by the name of Raymond to assist with the 60 PDs and one elderly, newly calved cow that had gone lame. Having completed the marathon fertility clinic, Cameron received an emergency call and abandoned us to the well-intentioned care of Raymond. I suppose the best way to describe him was cocky with a capital C. He knew it all and, when I showed him to the pen where the lame cow and her calf were being kept, he said those fatal last words, 'well, this looks straightforward enough.'

I explained to him that she had fallen lame about three weeks before her eventual calving date, and although I had inspected her thoroughly there did not appear to be anything the matter. After calving it had got worse and now the old girl could hardly move.

Raymond threw open the steel gate to the pen with an almighty clatter and then said in an ebullient manner, 'well, my girl, I've come to sort you out.' He took three bustling strides into the pen, which made his waterproof coveralls slap loudly against his wellies.

The cow's ears pricked at the noise, and although this crippled creature had her back to the vet, and to all intents and purposes was temporarily disabled, I instinctively knew what was about to happen. With a roaring bellow, the cow leapt to her feet, span round, and, making sure that she was between Raymond and her calf, put her head down and charged. Non-farming people will find it hard to believe the speed a cow can accelerate to in these circumstances; stationary one moment, at full tilt the next, and even if they have no horns, their heads are like sledge hammers. Raymond managed to sidestep her and she went crashing into the gated end of the pen. She gathered herself back together, pirouetted on the spot and immediately saw that Raymond was now between her and her calf.

People have died in this situation.

The young vet ran at the six-foot high side wall. Jumping up, he got his hands over the top of the breeze blocks and he managed to pull himself far enough up before the cow slammed into the wall beneath him, hard enough to sway the whole wall and crack two blocks. He clambered down the other side of the wall and I took him into my house for a cup of hot, sweet tea. He was pale and had that clammy look people have when they have had a fright. There was not a lot of conversation at the kitchen table. All his previous bravado had been an act, and clearly a now very sheepish young man was thinking whether he really wanted this for the rest of his life. His decisive answer came as he was draining the dregs of his cuppa, when I said, 'she will have settled down now, the poor thing. We'll go and have another go, shall we?'

With that he got up from the table and, without a word, went outside, got into his car and drove away. That was the last we saw of him. Apparently he handed his notice in to Cameron that afternoon, and that was the last he saw of him, too.

Farmers tend to socialise with other farmers, and because farm life is so busy the party calendar tends to be a bit sporadic. I could not have predicted how my social life and circle of friends would explode exponentially with my introduction to the Round Table organisation. This occurred when one of the two Woolaway prefabricated bungalows that were opposite the farmhouse became vacant. The farm tractor foreman we had recently hired lived locally and didn't require accommodation, so we rented the house to a couple who were having a house built in Huddersfield. Their names were Ron and Claire Sorkin. Ron was the newly appointed managing director of the Rug Kits division of the Redicut Wool Company. It was based in Wakefield, which was a centre of excellence for wool production, spinning and carpet production. Ron and Claire were London city types through and through and living on a farm was a new and different world for them. Ron was quite comfortable with his new surroundings but Claire struggled with all the open space and the smells.

We always had a problem with vermin at Manor Farm. We had a good number of farm cats to try and keep the numbers down, but the constant deliveries of regular feed, as well as the cornflakes, chocolate, etc., which were kept in open-ended bunkers made it impossible to eradicate the problem. About once a month, one of the dairymen would bring his ferrets and Jack Russell terriers in and we would have a culling session. At the end of one of these, we lost a ferret and after searching and calling for him, we gave him up for lost.

About mid morning the next day, there was a frantic banging on my front door and when I opened it, Claire threw herself into my arms, and in between her hysterical sobs told me there was a wild creature, a hairy snake at least three feet long outside her door, smashing her milk bottles. She was so frightened she had climbed out a window on the other side of the house and dropped six feet to the ground and sprinted to my front door.

My mind raced – what could it be? A badger? A fox? I went to investigate, and found Arnold, our lost ferret playing with empty milk bottles by Claire's front door. For a ferret, Arnold was very friendly and quite cute. Unfortunately, my idea of cute and Claire's were two entirely different things. When I had put Arnold safely into the plastic cat carrier we kept in our garage, I took Claire back into my kitchen and gave her a cup of coffee. She was very upset and poured her heart out to me, saying how all the open space, the lack of street lights, how long it took to get to the shops etc upset her. She remained very upset until Ron came home and took over the manly protector role.

The next week Ron came over to see me while I was milking, and by way of thanking me for my chivalrous behaviour, invited me to a whisky tasting with Pendragon Round Table. I had no idea what sort of club the Round Table was, but whisky tasting sounded like a good idea. The following evening I set off for the Golden Cock, a very traditional stone built pub at Farnley Tyas. Ron was going directly from his work and we were to meet in the bar at 7.30.

When I arrived I parked my Land Rover in the car park at the back. I didn't have a clue what to expect and I suppose I was a bit nervous, and as I opened the door into the bar a blast of noise greeted me. Inside were about thirty-five men of all shapes and sizes, all having a good time. The only thing they appeared to have in common was that they were all under the age of forty (I learned later that in the Round Table, when you reach the age of 40 you're out). Ron was at the bar and after buying me a pint we turned and joined the mêlée. He introduced me to everyone, which was all very nice but the instant bonhomie and camaraderie made me feel uncomfortable. Although I stayed a member of the Round Table over the next sixteen years and was a member of five clubs the length and breadth of England, I never got used to this instant friendship, but in amongst the extrovert rabble, I would find lifelong friends.

At 8 o'clock we all went upstairs to a function room and sat at one long rectangular table to eat a three-course meal with wine. When coffee was being served the chairman carried out a bit of routine club business before introducing the evening speaker, who came from the Scotch Whisky Association. He placed 10 bottles of Johnny Walker's along the centre of the table and gave us all an engraved crystal whisky tumbler. His first instruction was we should rinse out the glass and our palates with a shot of blended. 'No spittoons here,' he said in his Highland brogue. 'It's noo reet to waste any o' the nectar.'

When all our palates were cleansed, he gave us each a generous measure of the first single malt of the evening. After we had expressed our praise for the elixir he directed us to the cleansing effect of another shot of the Johnny Walker, then he gave us particularly interesting, peaty flavoured malt from the Isle of Islay. I am not at all certain how many malts with blended chasers I drank that evening. I lost count at ten, I think. We left the Cock at well after midnight, and much to our communal shame we drove home. How I managed to do that is a complete mystery, but we were all so drunk no one had the common sense to call a taxi.

The next day found me with a very thick head and jelly sticks for legs, chasing cows out from around the farm irrigation lake. Fishermen from the International Tractor Company fishing club (International had just bought the tractor business from David Brown, who, you will remember had owned Manor Farm before the Coles Brothers bought it) had left the gate open again. Leaving gates open is symptomatic of urbanites' knowledge of the countryside code and the perception they can go anywhere with impunity. This situation at the lake was an on-going problem and there was an historical legacy behind it from the days of David Brown's ownership. The fishing club stocked and managed our body of water, keeping the banks trimmed, the reed beds contained and the fish stocks healthy.

When I got back to my office, I rang the factory up and asked to speak to the person in charge of the fishing club. 'That will be Mr Davis,' said the efficient-sounding receptionist. 'I'll put you through.'

Wanting some idea of how to pitch my complaint, I asked whether Mr Davis was just in charge of recreation or something else as well.

'He's the Director of Labour Relations, Mr Dixon,' she replied. 'Sport falls under his remit. I'm putting you through now.'

I wasn't expecting to speak to such a high flyer and I did some quick thinking. After all, this was a multi-million pound company, International by name and international by nature.

'DAVIS!' The name was barked down the receiver at me. How he managed to achieve in one word both his accent, straight from the Welsh valleys and such aggressive hostility, is beyond my comprehension.

'Good morning Mr Davis,' I said. 'My name is Philip Dixon. I'm the Farm Manager of Manor Farm and we are

experiencing some difficulties with the factory fishing club.
I was wondering if you would be so kind as to pop up and
have...' I never managed to finish inviting him to coffee and
discussions.

'"POP UP?" I DON'T "POP UP" ANYWHERE!' With that he
slammed the phone down and that was the last I heard from
him.

I was furious to be spoken to in that way, especially
as they paid no bye for the use of the lake; we were just
honouring an old tradition. I locked the gates to the lake and
put Swinney into the field. When you put a dairy bull into a
field you have to put a notice up saying, "Beware of the Bull."

I added, 'If any fishing club member wishes to use the
lake, would they first kindly ask Mr Davis to open some sort of
discussion with me at the farmhouse. If not, the bull stays.'

No one ever came to see me about it and I never
heard from the man who was clearly a natural born union
negotiator. Not. If that was the calibre of our industry
captains, then no wonder we had so many strikes around
that time.

When I started managing Manor Farm I was fortunate to
be taking over a full complement of staff, so I was able to
learn interviewing techniques slowly over the months when
natural movement of staff threw up positions for me to fill.
Unemployment in the area was about 25 per cent, so there
were always plenty of applicants. Tractor men were always
the easiest to find, because they had acquired some of
the skills they needed in other industries. However, good
stockmen and dairymen, even at this time, were becoming
thin on the ground. One of my herdsmen had handed in his
notice because he was emigrating to Australia, so I placed an
advert in the Farmers' Guardian for a replacement. I had two
applicants. One was a young man, fit and freshly qualified.
He was full of enthusiasm but was from a non-farming
background and was relatively inexperienced. He reminded
me of me a few years earlier, full of youthful audacity.

The second was Maurice Wild. He was 58 years old and
had 58 years of experience in farming. In fact, he had many
more, because his father, Ken Wild, was still farming at the
age of 82. He had passed his knowledge on to his son, as
had his father before him. Life for Maurice at his home farm

had become impossible. If any of you have ever crossed the Pennines on the M62, you will have come across a section high on Marsden Moor where the carriageways of the motorway split to make way for a farm steading (the truckers call it "the little house on the prairie"). That, sandwiched between two screaming torrents of traffic, thundering east and west over that austere landscape, was Stott Hall Farm. It had been in the care of the Wild family for many generations and Ken Wild was the current tenant. Why the landowner, the motorway builders, the planners (the farm house was built in the eighteenth century and should have been listed), the council and the government couldn't have put their differences aside and built the road entirely on one side or the other of the farm is beyond my imagination. The official line was that the geological ground survey of Marsden Moor dictated the ignominious decision for the route of the new motorway.

Maurice had a different tale to tell, which, according to the M62 web site is an urban myth now. Ken Wild, although not the owner, had earned strong tenant rights by his family's length of tenure; it was his home and he refused to move. The powers that be, in an official tantrum, decided to punish him by boxing him in. No doubt they thought they were very clever, but as a consequence, Maurice was leaving home for the first time in 58 years and was in my office asking for a job, because the farm was no longer tenable for Ken and his extended family.

Although Stott Hall Farm was a sheep farm, Maurice had run a small dairy herd to augment the family's income. Clearly he had vast experience of farming, but did he love his cows? He had no references apart from a letter from his dad saying what a good reliable chap he was. That left me to decide between the younger man's academic training and Maurice Wild's lifetime of experience. The factor that clinched my decision was of a practical nature; if a dairyman has hands in good condition it reflects on how he looks after the cow's udders. If his hands are rough and calloused, then I'll bet a pound to a penny the udders they care for will be in a poor state. Maurice had hands the size of grain shovels but they were clean and well cared for. The younger man had black, oily, cracked paws.

I chose Maurice, and it turned out to be one of my better decisions. He had a wealth of knowledge on stock issues and taught us all some of the more ancient cattle arts. One in particular that springs to mind was that of casting a cow. I had read about this technique, which involves getting a cow to lie down, but had never seen it done. It was one of those things you don't believe possible until you see it happening in front of you. Maurice and I were across the other side of the farm looking at a group of dry cows (a cow's year is divided into two parts. For nine months she will be in milk, but for the remaining three she will be resting and recuperating prior to calving her next calf and starting her next lactation. When a cow is in this three-month period she is said to be "dry").

One cow in that group had managed to get barbed wire wrapped around a front and back leg. To take her back to the steading where all the cattle handling facilities were would have caused her more pain and suffering. As she had been moving around she had pulled the wire so tight that it was cutting into both her front and back leg, making them bleed profusely. We needed to get her to lie down so we could unravel the wire.

Maurice suggested straight away that we cast her. Sceptically, I agreed, so using my Land Rover we corralled her against the two walls in the corner of the field. Then, with a bit of calm handling, we managed to put a rope halter on her head. Maurice pulled a dirty old piece of rope from the floor of the Landy and gave it a good shake to get rid of the dust and straw clinging to it. As I held her head, he fastened a fixed loop around her neck, then passed the rope down her spine, holding it in position with two hitches, one just behind the front legs, with the rope circling her chest and the other just in front of the back legs, with the rope passing under her belly, making sure it was in such a position that it would not damage her udders. He then went behind her and pulled gently on the length of rope, ensuring the two hitch knots pressed down on one side of the spine. As he continued pulling on the rope she bent her legs and slowly sank to the ground. As she lay there I folded her head back along her body to make sure she didn't try to get back up. It was so easy it was unbelievable, like something Crocodile Dundee would have done. We removed the barbed wire with wire cutters and pliers and dressed her wounds with antibiotic powder, and then we went off to find out how it had happened. It didn't take us long. Some bastard townie had fly tipped a load of garden fencing material just inside the field gate.

Sometimes, of course, with some people you can get it so wrong. You assess people and make judgements, using your self-perceived skills. A young apprentice I employed as a trainee dairyman appeared to be well qualified, but he always got up late and was lackadaisical in his work. All teenage boys have a tendency to be task orientated and don't think outside the box. The other problem is that they lack the physical strength of a man, and despite what they may think, they don't achieve muscular stamina until they are at least 23 or 24, (It's interesting to note that in France, an agricultural nation, parents are responsible for their children until the age of 25).

I made allowances for my young apprentice and he just took it as a licence to be worse. One morning I absolutely bawled at him. The next day he was late again. I had no alternative but to start the disciplinary procedure. This involved three verbal warnings, which would have to be recorded in the discipline book, then three written warnings. If this failed to get him to improve, he would be dismissed. We ended up following it all the way through.

The day of his dismissal I received a tearful telephone call from his mother, begging me to give him a second chance. I brought him into the office again and sat him down. He said to me he just couldn't wake up in the mornings because his head felt so heavy. I told him to go and see a doctor, but if he were late again I would be into his bedroom with a bucket of cold water. He went to see the doctor that night and never returned. I stood at his graveside four weeks later. He'd had a massive brain tumour that was inoperable. If I had realised there was something seriously wrong I could have got him to a doctor sooner. And I would have liked to have known if treatment would have been possible, but I was too much of a coward to ask. How could I have missed the signs? My self-perceived skills had let me down badly.

I was a keen member of the West Yorkshire Grassland Society and attended monthly meetings, although only in the winter months, as during the summer I was too busy. I won't bore you with any of what went on at these meetings, as they are rather technical in nature, but the talk revolves around the fact that if you grow grass you need to do something with it. With us farmers, almost all the grass we grow is fed to farm animals. When these animals have eaten vast quantities of this green fine dining, they inevitably produce larger than usual quantities of farmyard manure, or as we call it, FYM. This provides the grassland farmer with a good fertiliser for the following year's crop of grass, hay and silage. Thus it has a value, and a good deal of farmers' time is taken up with literally talking shit!

Most dairy farms in Britain handle their FYM in the liquid form, which is called slurry. This is stored in lagoons or in large above-ground steel tanks, waiting to be spread on the land in the winter. The main problem with storing it in bulk is that over time it separates into a crusty top layer and a runny bottom, so to speak. The recombining of these two layers into one homogenous gloop takes money, time and effort and was a favourite topic of conversation for the members of the society. Indeed it got so serious, that at one point we arranged various visits by manufacturers who provided demonstrations of their equipment, so that we could find workable solutions.

The first of these took place at the farm of a close neighbour of mine who had recently had his cow housing renovated. He'd had the slurry lagoon built directly beneath the cows, allowing the offensive material to drop straight in, via preformed concrete slats which were set at predetermined intervals. The only entrance to the slurry pit was via a large concrete trap door, which opened upwards through 180 degrees, so it lay on its back, flat on the floor when opened. It was large enough for all thirty members of our club to stand around, and the pit was full, with the crust flush with the floor, so it was impossible to distinguish between *terra firma* and *terra not-so-firma*.

Our speaker was just making his introductory preamble when a late arrival bustled into our throng. Because he was what we now have to call "vertically challenged," he pushed his way to the front. He saw this free space in the middle, and before any of us could stop him he stepped onto the crusty part of the floor. He disappeared in a second into the slurry. If he weren't to reappear quickly in exactly the same place he had broken through the crust, he would drown, and there would be no chance for anyone to break a new hole for him to resurface through. Fortunately for him, a few worrying seconds later, he slowly rose to the surface looking like some monster from the bog. We dragged him out; he was covered head to toe in an oozing six-inch layer of cow muck. And he stank. As we realised he was unharmed, the sniggering started, then the giggling, and finally, the side-splitting laughter. From that day onwards he was known as "Shit Boots," even when he became the chairman of the club and had official functions to attend.

The next demo was at a farm where a Howard Harvestore system had been installed. You may have noticed them as you drive around the countryside. They consist of dark blue tanks about four yards high and about forty yards in diameter. They suffered terribly over the years with the build up of slurry crust and some machinery company had invented a system to deal with the problem. It consisted of a powerful, tractor-driven pump that drew the liquid from the bottom of the store and blasted it through a nozzle into the top, acting as a cutting lance with which you could chop up the crust. It was automatically controlled by infrared sensors that directed the nozzle in a slow, criss-cross pattern over the whole surface.

It was a gloriously hot autumn day and the demonstrating firm had erected a scaffold walkway round the perimeter of the store. On arrival we were all directed to designated parking spaces and the club members threw open the doors of their vehicles so the interiors did not get overheated, and made their way to the gantry viewing platform. Normally there would be about thirty of us, but that day there must have been a good sixty. We were surprised, and the sales chap was delighted. After delivering the inevitable verbal diarrhoea that salesmen are renowned for, he fired up the tractor that was to drive the machine and took out a bullhorn so he could continue his description of his marvellous machine over the noise of the tractor. The business end of the machine was emitting a jet of slurry almost powerful enough to cut through concrete. He began to describe the progress of the jet as it started from one side of the surface of the slurry, but what he failed to realise was that the movement of the jet had nothing to do with the infrared controls. It was because the pressure in the pipe was so strong it was straightening the goose-necked shaped pipe.

By the time the slurry lance had reached the far side of the lagoon, the salesman had turned to his audience, with his back to the unfolding disaster. The jet of slurry continued its path across the surface, worked its way up the wall of the store and having cleared it, the arc it then formed quickly crossed the track and started to re-spray one of the lines of parked cars. Windows and boots had been left open, and there were two soft tops with their roofs down. Before the pump could be stopped, the whole line of Range Rovers, Saabs, Subarus, Land Rover 90s and Discoveries were a lovely shade of browny-green. Where the windows had been left open, they received a matching interior. The cost of that demonstration to the manufacturer's insurance company must have been enormous, and as you can imagine, the salesman didn't sell any machines that day.

The personal life of a dairy farmer is a curious mixture of home time and work time. A normal day (that's on the rare occasion a "normal" day occurs) is punctuated by routine tasks like milking or feeding stock. All meals occur in the house with the family. Sometimes they are rushed, and other times they are long, lazy feasts. Although my actual working hours were long, they were interspersed by large amounts of family time, so my days were amazingly fulfilling. A farmer is not the nine-to-five man, disappearing first thing in the morning never to be seen again until late evening. Having taken that decision to expand the family, we didn't have long to wait for a new baby to come along, and when it did I lost the help of Carol as my relief tractor driver/farm secretary/milkmaid/ general gopher. As her confinement drew on, we resolved I would be present at the birth, but little did we know just how present that would turn out to be.

One of the skills you acquire as a dairy farmer is that of midwife. All the cows give birth once a year and at Wilshaw that amounted to about 350 a year. Most animals manage on their own, but you try to be in attendance in the final stages, should they need your assistance. Over the years you face an extraordinary number of complications, most of which you take in your stride, so that you very rarely need to call the vet. I had missed Sharron's birth in Newcastle's Princess Mary maternity unit because a midwife told me she would be hours yet, and to come back in the morning, only to find that Sharron had arrived half an hour later. At Manor farm we lived too far away from the nearest maternity unit, so home deliveries were the order of the day for healthy souls in Wilshaw. Carol went into labour at 11 o'clock one evening and I phoned the midwife, who lived in Meltham, which was the nearest village, about ten minutes away. Wilshaw was just a hamlet. She asked me how long the time was between Carol's contractions. 'I'll give her a couple of hours,' she said when I told her. 'Then I'll be up to see how she is getting on.'

As before, in Newcastle, the baby was crowning within 30 minutes. I knew what to do, but the emotion had me sweating like a fat lass in a chip shop with no money. Ten minutes later I delivered Tracey-Emma. It was one of those occasions when time seemed to pass at two speeds. The birth dragged by second by second as Tracey-Emma inched, little by little into the world, but then when she arrived time seemed to flash past in an instant. I had another daughter.

At this point the midwife bustled into the room and I felt an enormous relief that the experts had arrived. Later on Carol and I received a severe dressing down for not getting in touch earlier, but our explanation was simply not believed.

It was *déjà vu* eighteen months later, when Gavin arrived. Having done this once you would have thought when Gavin put in his appearance I would have been prepared emotionally for his birth, but it was still enough to bring tears to a glass eye and turn your tummy inside out. They were beautiful moments in my life, of which I am very proud.

Upper House Farm and a xeroradiograph

Life was sweet. I was managing a large dairy farm in a beautiful part of the country and my family had been enlarged by the arrival of two more beautiful children, Tracey-Emma and Gavin, but I had not yet achieved my dream of being a farmer in my own right. This was never far from my thoughts, and I regularly sent off for the particulars of farms to rent. I applied for the tenancy for many, but even though I was frequently short-listed, my applications were always rejected. My problem was that I still lacked provenance. The letting agents were a conservative crowd and all had the business acumen of a bumblebee. As I was the son of a pharmacist with no close relative that farmed, all my qualifications and proven experience counted for nothing with them.

To buy a farm in England on agricultural wages was impossible, so being a tenant was the only feasible option. One morning, while I was worming a group of heifers in the handling pens, Carol came rushing out from the house, waving that morning's copy of the Yorkshire Post. She clambered over the five bar gate at the side of the cattle crush and thrust an advert under my nose. It was for a sixty-acre farm for rent or sale. The place was called Upper House Farm, and it was on the Dartmouth Estate, which was next to the neighbouring village of Slaithwaite (pronounced *Slowwit*). It was local, but more to the point, with a struggle we could afford it. I applied.

Becoming the tenant of a farm on the Dartmouth Estate was a bizarre process. Normally when you rented a farm, you rented it all, but with the Dartmouth Estate, you rented the fields and the land everything stood on, but you had to buy everything that stood on the land – the house, the dry stone walls, the buildings, the water troughs, even the trees.

This meant you had to raise money to buy the house etc, but the leasehold on the house, the permanent structures and suchlike was tied to an agricultural tenancy and most financial institutions hated this arrangement. Lloyds was the only bank who would finance these farms. There was also another unusual condition; Lord Dartmouth, the owner of the estate, insisted that a bank representative and the land agent jointly took the decision about whom the tenancy was awarded to. This was a situation in which I could shine, because the decision would be made on the ability and acumen of the applicant. I prepared my five-year business plan, a process I had learned at college but had become second nature in my present role. One afternoon in the estate office I gave an hour-long presentation. The following morning I received a telephone call saying I could proceed with taking up the tenancy. We were thrilled. It meant Carol could have a house she could call her own and the horses she had longed for, and I would be a proper farmer and not somebody's lackey.

Upper House Farm consisted of sixty acres of upland ground, one step up in quality from open moorland and was at the top of Cop Hill, above Slaithwaite. It was the genuine *Last of the Summer Wine* country, with Holmfirth, the village from where Compo and his mates roamed, just a stone's throw away. The house was a typical Yorkshire long house, built of Yorkshire stone with narrow mullioned windows and a Yorkshire slate roof. It was actually two small cottages combined into one. Ronald, one of the two brothers who previously owned the place, still lived in one of them. Attached to it was a hay barn with a large stone arched entrance. Next to this was a stone shippon, an old-style cow house where cows are tied up by the neck and milked. Above this was a large loft. Beside the shippon was a good-sized dairy with ample space for a bulk milk tank and an ancient bottling plant standing in a very heavily cobwebbed corner. The whole place had a view to die for, if the mist cleared long enough to see it.

Our plan was that I would keep on working at Manor Farm to pay the bank loan and we would start farming in a small way with a pedigree Charolais herd of beef cattle. Running a beef herd was our only option because the time restraints of my job prevented me from spending the time milking a dairy herd of my own. In my spare time I would begin renovating the house and preparing the farm for a more intensive beef project that would support us as a family, but it would take many years to build the herd numbers. It would be hard work, but it was a foot in the door and we were ecstatic.

The cottages were pretty dilapidated and needed modernising, because Ronald had never done any work on them. As the house was a listed building, the first thing we would need to do was secure the services of an architect and draw up plans. Huddersfield Polytechnic (now it's The University of Huddersfield), which was about ten miles away, had, and still has, an art and architectural department. I had the bright idea of contacting them and offering the design of my house as a final year project competition, with the prize being the industry standard charge for a piece of work of this type. The downside was I would have to put up with 15 students crawling all over the place, but the upside was that I would have a choice of 15 different designs that would all take into account all of the restrictions imposed by Kirklees District Authority. The senior lecturer asked my permission to do a preliminary survey. I readily agreed and we arranged to meet on site at 8.30am the following Thursday. I warned him that the old owner, Ronald, had not yet moved out, but I was sure he wouldn't mind him looking around.

The lecturer arrived and bounced out of his Citroen 2CV: a flamboyant vision in a fedora, camel coat and skin-tight velour bell-bottoms. Arms windmilling in the air, he bid us a 'good morning, I'm Gerald. Good. I'll just carry on shall I?' And with that, he disappeared into the bowels of the farmhouse.

After ten minutes I heard an almighty terrified howl, and went in through the front door to be greeted by Gerald sliding down the stairs on his backside, his pristine coat trailing behind him like a bridal train. At the bottom, he jumped up and, without looking where he was putting his feet, stepped into Ronald's gazunder, which was covered with a cloth, waiting for the contents to be disposed of. (There were no drains, septic tank nor cesspit here – Ronald went out every day with a spade to empty his chamber pot.) With his foot stuck fast, Gerald floundered around like a demented Mr Tumble. Eventually the ceramic pot could take no more and broke, sending its contents all over Gerald's lower half.

He was inconsolable. It took Ronald and me half an hour
to calm him down sufficiently so that he could tell us what
had spooked him in the first place. He said that while he was
in one of the back bedrooms, it suddenly went cold and
something fondled his bottom. Ronald said in reply that he
doubted it would have been the work of old Septimus.

'Who on earth is Septimus?' I asked.

'Ee,' said Ronald. 'T'owld ghost never bothers owt,
'cept a lass's bonny bum. He likes a bit of a feel, tha naas.'
(Translated, that means 'the old ghost never bothers
anything except a girl's pretty backside. He likes to feel them,
don't you know.')

After cleaning himself up, a not so happy Gerald climbed
gingerly into his Citroën and opened the sunshine roof to let
out the fumes of the chamber pot's contents. Then, in a very
distressed state he slowly drove away. After this, I thought
my deal with the college would fall through, but although
we never saw Gerald on the farm again, his class of students
did the business. Nine students entered, but only five sent
completed drawings. Three of them were far too outlandish
and the planning department would never have approved,
but the remaining two shared the prize, because we decided
to use bits of both designs.

As for Gerald's encounter with the other side, his gender
must have been very much in question if the house ghost
couldn't tell the difference. Or perhaps Septimus had,
and maybe still has a sense of humour. Septimus became
a member of the family and his presence was never
malevolent, and when we had finished the renovations
and created a large open plan kitchen, he often goosed
our female visitors, making them leap as they crossed the
kitchen. We never actually saw him, but the children always
blamed him for moving things we could not find, and
sometimes he even got the blame for eating the last slice
of cake or tasty treat Carol had put on one side for the next
day. Of any such acts I remained entirely innocent, despite
what the children might have said.

Considering the age of the house and its neglected
condition, renovations needed to go far beyond plaster
deep and the structure had to be tackled before we could
consider implementing the winning students' design ideas.
The front wall was bowed to such an extent that it gave the

house the surreal impression that it was pregnant. It was a supporting wall, so if we didn't straighten it very soon the roof would collapse. I bought a half dozen Acro props to support the upper floor, which would take the load off the heavy Yorkshire slate roof, but the moment the props took the weight of the roof, the front wall collapsed with a thunderous crash, sending a cloud of dust mushrooming into the road. I then realised that it hadn't been the wall supporting the roof, but the old oak beams that ran across the frontage, and the rubble filled wall had been held in place by the beam.

The plan had been to save the stone mullions and lintels after carefully dismantling the front wall, but the avalanche of sandstone and dry clay that had originally been used to stick the wall together had smashed them to bits. Replacing them was a job for a stonemason. The recession of the late 1970s hung on in rural Yorkshire right up till the mid-eighties and I was able to find an oldish chap who had been out of work for a number of years. He was called Edgar Batensby and he lived halfway down the hill to Slaithwaite, so he was practically a neighbour, but what's more he was able to start the next week. I had seen his work around on odd houses in the area and had been really impressed. I had intended to work alongside him for two reasons; one, to keep the cost to a minimum and two, to learn some of his trade by watching him at close quarters.

Edgar worked for me for just over a month and, in that time, rebuilt the wall using renovated mullions and giant stone lintels. He reset the roof and replaced some of the Yorkshire tiles, which are basically thin sandstone flagstones. He started by placing small tiles at the ridge, gradually increasing the size, so that the final row that met the gutters was made up of yard-square tiles. It makes an incredibly heavy roof, with the individual tiles hooked in place by mutton bones (these are what they sound like – real bones from sheep. I was told steel or copper fittings rotted out and did not last as long as mutton bones, which had the advantage of being free).

Edgar also re-pointed the whole building, including the barn. He put me in touch with his friend, a cabinetmaker called Bob Weatherspoon, who was also out of work. The pair of them worked together to fit the new oak window frames that Bob had skilfully crafted out of second hand timber that he had stored in his back yard. The bill for the job came to little more than two men's wages for a month. They appeared to have only charged a pittance for materials. I asked Edgar how the pair of them had managed to be so reasonably priced and he told me he had not had to buy anything. All the stone had been scavenged from the rubble from the long-abandoned Scammonden Mill, which for safety reasons had recently been demolished.

The month after the front wall had been rebuilt, it rained and rained. Upper House was built into the hillside with the rear windows being at the same level as the field, and the drainage ditch behind the house had long since filled with silt. The back wall performed well as a filter, because we ended up with a sparkling stream snaking its way across the room that was destined to be our kitchen, and out the front door. Judging by the groove the water had cut in the beaten earth floor this was not the first time this had happened. Sharron and Tracey-Emma thought it was great for playing indoor Poohsticks.

The ditch was a yard deep and fifty yards long and I had the monumental task of clearing it by hand. Back then mini diggers were not available for hire. The next problem was that I had no idea where to start, because I first needed to find the drain that emptied the gully, otherwise it would just fill up with water again, however much I dug out of it. I asked Ronald for ideas and he pointed at a spot at one end of the ditch, but when I tried to dig there I found, just under the surface, a not-so-dry drystone wall. So I stood at the top of the field and studied the way the land sloped down to the back of the house. At the point where the living accommodation ended and the barn began there was a clear dip in the land, and at the same point the stonework of the wall was slightly curved; almost vaulted. Here, I dug down a good yard, creating a pool of muddy water two feet deep, but probing with my spit (a ditching spade with a long handle and a pointed blade) I thought I could feel a large stone slab. I stuck the blade of the spit under one corner and pulled down, bending the all-steel spit in the middle.

All of a sudden, there was an almighty sucking sound and a vortex of muddy slurry disappeared so ferociously, I felt my leg may have disappeared into it as well if I had stepped into it. The water from the rest of the ditch rushed in to feed the hungry miniature whirlpool, enlarging an already huge hole

at the side of the slab. I had no idea where on earth all that
water was disappearing to, so I ran round to the other side
of the building and out into the field and about 100 yards
down, under an enormous tussock, I found a partly collapsed
ancient stone culvert, much like the old roman drains I had
been taught about at school. A torrent of brown soupy water
was spuming out of the end and across the field.

After the water that had been trapped in the drainage
ditch had emptied, I left the whole thing to dry out for a
couple of days, then cleared two trailer loads of debris out of
the ditch. The back wall of the house made up one side of the
ditch, whilst on the other, field side, as I discovered when I'd
cleared out the rubble was a beautifully constructed drystone
wall. Over the years that ensued after I had cleaned out
the ditch, this wall became covered in a fantastic variety of
mosses, ferns and lichens that would have graced any gold
medal stand at Chelsea Flower Show. After all this work, we
had no more flooding problems, no matter how hard it rained.

We carried on all that year, dismantling badly wormed
timber partition walls to create a large open plan area for
the kitchen and concreting over the beaten earth floors.
The plaster had long since disappeared off the walls, so I left
some of the sandstone exposed and re-plastered others.
When I say plastered, I do so with extreme poetic licence;
white stuff clagged onto a wall would be more appropriate.
A modern-day estate agent would describe it as "textured."

We had owned Upper House for just over a year and had
made the house watertight when disaster struck. Back at
Manor Farm I was inside one of the barns, servicing a Wilder
flail forage harvester, prior to the start of the silage season.
The harvester had a large funnel on the top of it that directed
the chopped grass via a hydraulically controlled spout into
a five-ton trailer that was towed behind a tractor. Since the
previous season the spout had seized solid, so with a pocket
full of spanners and a large mallet, I climbed twelve feet up a
ladder to fix it. Having hammered the 18mm ring spanner over
a seized bolt, I wrapped my left arm around the round chute
and with my right arm pulled on the spanner. While everything
was in place I was perfectly secure, but unfortunately
something gave way. Whether it was the bolt or the spanner
slipping, I will never know, but the back flip I then executed

was of Olympic standard. Unfortunately the landing was more like an over ripe peach falling from a tree, rather messy and totally unlike a medal-winning Lewis Smith dismount. My arm muscles were strong enough to lock and brace my arms, but unfortunately the bones in my left arm weren't as robust. My hand ended up where my elbow should be and the bones in between were just crushed into crumbs.

I have always prided myself on my manly pain threshold, but this really stung. As they wheeled me down the corridors of Huddersfield hospital I was wincing at the jolts from the sealed joints between the vinyl tiles on the floor. They sent me home twenty-four hours later, after they had pulled my swollen injured arm roughly into shape and wrapped it in a cotton wool lined plaster cast. A week later I went back in to have an operation to sort it out. I woke up the next day with a hazy sensation, the like of which I had not felt since the last Genesis concert I had been to, when it appeared the rest of the audience was smoking stuff they had extracted from little silver foil packets. The doctor put me on Omnopon, a morphine derivative, for the extreme pain, which I didn't feel until forty-eight hours later, when they switched me over to paracetamol. The operation had not gone well, and after five hours of failing to wire it all back together they stitched me back up, but told me nothing. They discharged me the following day and told me to come in to outpatients regularly until the scars healed.

In between hospital appointments I carried on managing the farm from the office, hiring a young man to perform the physical work. Obviously no progress was made with Upper House. Carol travelled over to Cop Hill on a daily basis to look after the small pedigree Charolais herd of beef cattle we had started, which were fairly self-sufficient and did not require the level of commitment a dairy herd requires.

After four months of medical treatment, including two operations, I found myself sitting opposite my consultant orthopaedic surgeon. 'Well, Mr Dixon,' he said, exhibiting the bedside manner of Genghis Kahn. 'There's nothing more we can do for you, We can't get that bloody arm of yours to work, so we'll just sign you off as fifty or sixty per cent disabled, shall we?'

I was too stunned to say anything; at no point did I ever dream they weren't going to fix it. 'What do I do now?' I said.

'You'll just have to reorganise your life,' was his caring reply.

I could not believe it. They had written me off. I had a left arm shorter than my right, and virtually no movement in the elbow. The nervous system that controlled hand movements was unpredictable, and my arm was made up of crushed bone that had not fused together. If I knocked my arm, or indeed tried to use it, the pain brought me to my knees. It was clear my days as a working farm manager were over if I didn't do something ...

Through Round Table contacts I managed to get an appointment for a second opinion at the Robert Hunt and Agnes Jones Orthopaedic Hospital in Shropshire. What a difference there was between the people there and those at the Royal Infirmary in Huddersfield. Here, they were so professional; so efficient. It must be said that within ten minutes of arriving, my "poor me with my broken arm" attitude was put into perspective. I felt an utter fraud; I was introduced to complete paraplegics who were fighting to move a finger, and laughing about it. They quickly sussed me out, saying things like, 'There's no arm in him,' or 'he's an 'armless chap.' One chap was working with Land Rover to work out how he could compete in some international four-wheel drive hill-climb competition using only his mouth. And it was a crash in a Land Rover that had put him into hospital in the first place! Here was courage I could only dream of.

I was told by a very pretty nurse to get undressed, put on the gown on the end of my bed; if I needed any assistance I was only to ask (Oh, get thee behind me, Satan). All gowned up, I lay on my bed and watched as the consultant paraded down the centre of the ward with his entourage, stopping intermittently at beds like some sort of superstar with a swarm of hangers-on seeing to his every need. Eventually, they gathered around my bed and there, before my very eyes, was the smiling, bearded face of the twin of James Robertson Justice, the star of the *Doctor* comedy films of the 1950s and 1960s.

'Good afternoon Mr Dixon,' he said. 'My name is Mr Williams and this bevy of white-coats are my students. I hope you will not be offended if I ask one of them to examine you – under my supervision, of course?' I nodded acquiescence, and what followed was very much like a scene from one of those films.

'Tarquin, dear boy,' he said to one of his entourage.
'I wonder if you would examine the extent of Mr Dixon's injuries?'

Tarquin looked like what you'd expect a Tarquin to look like: thin, pale and weedy but at the same time having a supercilious air. This young emaciate took my injured hand in his wet lettuce of a grip and asked me to squeeze it as hard as I could. I obliged, although doing so hurt me a bit. Tarquin seemed very pleased with himself, commenting to the consultant that, considering the extent of my injuries, I had a good strong grip.

This wasn't enough for Mr Williams. 'But what is the extent of loss of strength?' He enquired.

'There is no way,' Tarquin replied, 'of knowing what it was before.'

Williams studied the young man. 'Don't most people have two arms, boy?' He asked. 'And under normal circumstances don't they have approximately the same strength?

Tarquin began to look sheepish.

'Compare it with his right hand, man.' Mr Williams suggested.

As a deflated Tarquin moved round to my right side and bent to his task, Mr Williams' smiling, bearded face appeared over his shoulder. 'Mr Dixon,' he said. 'Would you give the young doctor's hand a good squeeze?' And then he winked at me.

I squeezed as hard as I could.

Tarquin yelled and fell to his knees.

A bit of difference, is there, young man?' Said Williams to Tarquin. Then, to me he said, 'Thank you, Mr Dixon. Very instructional.' With that, he turned his back and chuckled all the way down the ward, muttering, 'he'll not forget that in a hurry!'

There were two tests they performed at the hospital that I remember in detail. The first was a xeroradiograph, which was a sharp-focused coloured picture of the bone structure (or rather, the lack of it) within my arm. When I saw this, I fully understood for the first time the hopelessness of my expecting any natural recovery. The second was a test of the motor nervous system. This consisted of taking a rather large needle with a miniature microphone at the tip, inserting it first into the back of my hand between my thumb and index finger, then into the back of the muscle at the base of my thumb. The sensations in my hand had reduced after the accident, so despite the size of the needle it didn't actually hurt much, but coupled with the roaring and scraping sounds from the loudspeakers as the medics moved the microphone into position, it gave the impression it should be life-threatening.

At this point I need to confess I hate electric shocks. Years back, Mr Hall at North Farm would grab my arm in his vice-like grip, thinking he was being amusing and then grab an electric fence, so the pulsing current would pass into me. Then he wouldn't let go. I will walk miles now to switch a fence off, thus avoiding touching one, and I use a meter to test if it is working, and not use a blade of grass, which most tough farmers do, saying it just gives you a tingling sensation. I've tried it. It gives you a nasty shock. So, what did the good Doctor do to me next? He attached an electrode to my shoulder and gave me shocks, which contracted my thumb muscle, which in turn squeezed the microphone. The sound gave them a precise measure of the functionality of my internal wiring. I've had better moments in my life.

At the end of my stay, Mr Williams and I had a long conversation about my options. They were: one, have my lower arm removed and a plastic arm and a hook fitted; two, wait until the bone mass fused together, then return for a series of procedures to give me a plastic elbow joint that would only be capable of very light duties. This would take two years to complete. The third option was to do nothing, which cosmetically was the best solution, but effectively I would be living with a broken arm, which would probably be painful for an awfully long time.

My vanity chose the third and I returned to Manor Farm to face the fact I could no longer function as a working farm manager and I had to make a go of Upper House as a source of revenue. My instinct as a hunter-gatherer seemed to have heightened exponentially as my self-image of masculinity diminished. I felt I was not a whole man. So with a wife and three young children under six years old, I was dismissed from my job at Manor Farm. We sold our Charolais herd and bought some elderly Friesian cows and we moved into Upper House Farm. I resolved that if a bunch of ancient Egyptians can build pyramids with no machinery, then a one armed Geordie could make a go of it as a farmer.

Just before I sold our Charolais, I sat Sharron on Snowy, our White Charolais bull and paraded her around the field like a champion jockey at the Derby in the winning enclosure. I had fulfilled a dream. I had embarked on another life adventure and become a proper self-employed farmer. It was scary and exciting at the same time.

Rebuilding works, fireworks, and duck races

Second only to a watertight roof and somewhere comfortable to lay your head, getting clean, especially on a farm, is of paramount importance. Unfortunately, there was no sewage system of any sort at Upper House, so putting one in was necessary before we could install any running water in the house. The cows and the dairy were already provided for; looking after the animals first is fairly typical of farms worldwide, but waiting for proper bathroom facilities when you have three young, farm-dirty adventurers to get clean was not an option.

In the dairy was a large D-shaped plastic trough on metal legs, which had hot and cold water feeding it. At the end of the day, when the cows had been fed and milked, I would fill the trough with hot water, go back into the house, wrap our three grubby children up one at a time in a large fluffy bath sheet and carry them through the barn, which was filled with the compressed sunshine smell of new hay, through the shippon, between the rows of sleek, gently grunting cows that never even bothered to raise their heavy eyelids, and into the dairy.

The cold night air in the dairy was full of steam that rose from the hot water in the trough. Sharron, being the oldest, was first in and last out. Tracey-Emma was next, and she would cling to me for dear life in case I dropped her. It never happened, but it was always a possibility because my gammy arm would suddenly go limp with no warning. Last in was baby Gavin. By the time he was in the trough, the water would be coming over the top and onto the floor. It was a riot of splashing and having fun. When they were all well scrubbed I wrapped them up tightly in their towels and took them back to the house. The girls dried and pyjamered themselves while I dried wriggle-monster Gavin. The house

was not centrally heated and tended to be a bit chilly, so
Gavin wore a teddy bear style baby grow to keep him warm.
Try putting one of those on an excited, giggling child when
you've only got the use of one hand. It's a real challenge.

Getting a septic tank up and running was essential and,
because money was tight, I fitted it myself. One section
was a large plastic tank the shape of a prize-winning onion.
Installing that was easy. I hired a JCB, dug a hole, lowered
it in and then flooded the rest of the space with 6 cubic
metres of Readymix concrete. The soak away, made up of
200kg preformed concrete slabs, was a different prospect
altogether.

One Saturday morning I was at the bottom of the pit. I
had one of the side panels suspended from the rear loader
of my ancient John Deere tractor that I was trying in vain
to position. I was virtually screaming at the frustration of not
being able to perform normally. I looked up and twenty
heads appeared around the perimeter of the hole, all
laughing at my pathetic impersonation of Mr Angry. It was
the Pendragon Round Tablers, who had arrived *en masse*
to help me out. Most were accountants, solicitors, doctors,
office managers etc, who wouldn't know a spade from a
shovel, but over the ensuing weekends, until the house was
fully habitable, they continued to turn up. Me being me, I
hated to think of myself as a charity case, but these were my
friends and we had a ball working together. If the tables had
been turned would I have helped any of them? I would have
been first in the queue. What sort of mixed up logic is that?

The day we got a fully functional bathroom was a real red-
letter day, or to be more exact a browny-purple day. I had
bought a top-of-the-range, ex-showroom Villeroy and Boch
suite, which included an enormous corner bath. It's only
problem was its colour, which was a muddy purply, browny,
reddy sort of colour, an extreme colour even though it was
probably made in the psychedelic 1970s and already at least
10 years old, but that was why it was so cheap. As Ronald,
the house's previous owner, would say, 'nowt's for nowt.' He
thought I was completely mad spending time and money on
our living conditions when there was proper work to do out
on the farm.

Ronald was a proud Yorkshire farmer, born and bred.

He was a dyed in the wool bachelor. His brother, Edwin, a worldly married man who had once been to Leeds, had kept a brotherly eye on him, Edwin's wife cooked Ronald's meals for him and the couple took him in after we bought the farm. Ronald seemed to be about 150 years old and had a wealth of knowledge inversely proportionate to the condition of his body. All that was left of him was sinew and bone. He looked like a desiccated Popeye. He wore woollen clothes that had been clean twenty years ago, when they probably would have fitted him. His flat cap was superglued to his head, and there it remained, even when he was eating his meals. I think he might have worn the cap to bed, too. But little did we know, when we exchanged contracts for the land, the stone walls, the barns and the houses, the deal also included this overgrown gremlin, who grumbled at every change we made.

Every morning at 7 o'clock sharp I would look out across the fields and there Ronald would be, wending his way to me, through rain, fog, and more rain. The first job of the day for him was to feed the hens and collect their eggs. In the field behind the house was what looked like a miniature railway truck on wheels. This was the hen house, or cree, as it was known in Yorkshire. Inside were our hens' nesting boxes and their nightly roosts. Ronald would open the door so that the twelve-strong brood could run free for the day, and his last job of the day was to shut our poultry away safely in the cree. This was a responsibility he took very seriously.

One morning, a year and a half into this routine, I had almost finished milking when he arrived in the byre, his ruddy face redder than normal and his larger eye rheumier than ever. 'A'm away t' t'ouse fur t'gun,' he growled, and then promptly disappeared like some magical creature from the forbidden forest.

It took a few moments for the news to sink in that Ronald, who squinted because of his poor eyesight and had hands that shook with the onset of Parkinson's disease was going home to get a gun. As quickly as I could I stopped what I was doing and went outside, to see Ronald disappearing into the mist. After an anxious wait, he re-emerged, presenting the most forbidding sight. You know that at the beginning of the film Gremlins, the creatures are cute and cuddly, but half-way through, they turn into nightmarish ghouls? Well, imagine

139

one of these ghouls armed with a shotgun and you will have a good idea what was emerging from the abyss.

I called to him to find out what on earth he was up to and why he needed a double-barrelled shotgun. He didn't reply. He cut across in front of me, bulldozing his way to the hen cree, shotgun held at hip height like John Wayne entering a saloon full of bad guys. He jerked open the door, and because the door opened towards him and the entrance ramp was behind him, he lost his footing and started to tumble backwards. One barrel of the gun exploded into the early morning silence, punching a hole up through the felt covered roof and sending chunks of timber and bitumen into the air like the spume of ash from a volcano. The thought of the madmen who lose control and murder whole communities flashed across my mind as Ronald continued his inevitable backward trajectory, propelled by the recoil of the shotgun.

Suddenly, from the door of the hen house a red blur emerged. Seeing its way blocked by a gun-toting gremlin, it launched itself into the air as if it were Red Rum at Beecher's Brook. I had just focussed enough to see it was a large dog fox, when the second barrel of the gun thundered into the air and the beast tumbled to the ground, dead. Ronald continued rolling and as soon as his back made contact with the ground, and he rolled sideways and backwards. Drawing his knees up, he was on his feet in seconds as if the whole manoeuvre had been deliberate. Clint Eastwood would have been proud of the shot and Lewis Smith would have been delighted with the gymnastics.

Once Ronald was on his feet, he delivered his lecture on foxes. They were, he said, the foulest creatures that roamed the earth. The only good fox was a dead one. They were the serial killers of the animal kingdom; they would kill every hen in the shed and take only one to eat. Sure enough, the creature had killed all our hens. It had tunnelled its way in through the bottom of the hut and torn away the timbers. He'd then wreaked so much havoc on the inside that he had blocked his own escape route, hence his demise. The bastard got what he deserved.

Despite Ronald's objections, I continued to work on Upper House. In the kitchen I managed to fit an anthracite-fuelled

Aga I had found at our local scrap yard. These highly prized cookers are ridiculously simple, but as well as being incredibly heavy they are works of precision engineering and need the skill of a toolmaker to put them together. Unfortunately I had got something wrong when I assembled this one, because when we had strong gales the hot plate used to glow red. My father, known by now as Grandpa Dixon, found out quite how hot it got, when, one day, as he stumbled, he put his hand onto the hot plate to save his fall. There was a distinct aroma of burning flesh as his hand sizzled. The burn was so severe that it took several months to heal.

Upper House's kitchen was enormous – open plan, as one of the young designers had called it. The whole of the ground floor, apart from one small corner snug room, was used, and the enormous cracked and knurled oak timbers holding up the upper floor just cleared 7 feet, giving an even greater impression of floor space. Conventional sized tables looked lost in it, so I embarked on a quest to find a proper farmhouse table, the sort of thing where Enid Blyton's Famous Five would have eaten their ham sandwiches and drunk lashings and lashings of ginger beer. I didn't mind what sort of wood it might be, and although a rectangular table would be best, I was prepared to look at oval or round ones.

Affordable large tables seemed to be like hen's teeth (they don't exist) but eventually, after a six-month search, an ad in the Yorkshire Post came up trumps when I found a four-metre long, one metre wide pine Suffolk baker's table for sale. As the ad said to contact Yorkshire Television Graphics Department, I was convinced it was one of Emmerdale's props. It sounded ideal, although it must be said I did not have a clue what any sort of baker's table looked like, let alone a Suffolk one. After a quick telephone call and having received directions, I found myself winging my way, under a swirling cauldron of dark clouds, to Ilkley Moor. The directions took me to an area of the moor where you would not have been surprised to see Heathcliff and Cathy cavorting. A single-track road wound its way up to a dimly lit ancient stone chapel, which had been bastardised by knocking enormous picture windows into the upper floor (I even think TV house rebuilder Kevin McCloud would have winced).

I swung the enormous sanctuary-style knocker on the church door, and after a few moments, there in front of me was a tiny, very shapely woman, with a waterfall of shiny black hair, long enough for her to sit on. She was dressed in a long, close-fitting tartan skirt and a low-cut, clinging bright red top that showed a large amount of her ample cleavage. She was hot, in a vampish sort of way, until that is, she opened her mouth. 'Darling,' she said, windmilling her arms around like some sort of actress in a Charlie Chaplin film. 'Do come in. You've come to see me about my table!'

She sounded like Fenella Fielding playing a seductress in one of the *Carry On* films. She should have been sexy but somehow missed the mark. Pride oozed out of this red-clad siren as she showed me around her upside-down house. The bedrooms were downstairs and the loft-style living space was upstairs, with the large windows framing the view over the dark, satanic moor. The inside was very stylishly decorated and furnished, and looked much better than the outside. She was particularly proud of her bedroom and made sure I was aware that she wouldn't mind sharing it with me.

The table was upstairs, sitting perfectly in front of a large window. It was exactly what I was looking for. The whole of the tabletop hinged up to reveal a shallow storage space that ran the whole length of the table. She had used it for storing her designs, but originally it had been where the baker would prove the dough before baking. I asked her why she was selling it and she told me she had a little house in Tuscany and flats in Bath and New York. This converted chapel, which she described as a "darling house" was just too much.

We agreed a price and I came back a few days later with a tractor and long bale trailer and Don, one of my neighbours. They had put a pair of French windows at one end of the open plan upper storey, through which Don and I were able to haul the table and lower it onto the trailer. Later that month I managed to buy two old pews at a saleroom to put either side of it. I could sit sixteen people at it without a squeeze. I loved it.

When my Round Table work force was assisting me at weekends, our lunchtimes would become very French, with the meals, held around my baker's table often lasting two or more hours. We'd carefully transport several 2.5-gallon buckets of real ale down from the Rose and Crown to help wash it down. One of our team was a detective in the local constabulary and a debate ensued about whether low levels of alcohol improved one's awareness of the surroundings. Our police friend was horrified at this concept, and volunteered to bring his drive reaction test kit the following weekend to prove to us what rubbish we were talking.

The test kit consisted of a set of car pedals and a series of different-coloured lights, which were switched on by pressing the accelerator pedal. When one of these turned to red, the operator hit the brake and clutch pedal as fast as possible. The reaction time came up on a digital display that registered to hundredths of a second. We decided that it was a competition, although our friendly policeman became irritated at us treating it lightly. He said it was serious, and what we learned here today could well save lives in the future. Nevertheless, we decided that, at midday, we would all take the test, then immediately afterwards drink a pint of beer. The next test would be at 12.30, followed by another pint. We'd have a third pint at 1 o'clock, a fourth at 1.30, and finish with the final test at 2pm. Then we'd go back to work. Reluctantly, we all agreed to this onerous task.

Much to our surprise – and to the utter dismay of our policeman – we got quicker, to a man, the more beer we consumed, some of us being 25 per cent better at the end than we were at the beginning. What he found utterly confusing was that it was true for him as well. Of course, he was right that we should not introduce an element of competition in the first place, because this induced an adrenaline rush. Practice played a part, too; the more we did it, the easier it got. Still, it is a little perplexing when you consider the weight of evidence we have today is to the contrary. I must say, I am in no way advocating drinking and driving.

House renovations generate loads of rubble and bonfire material. The hardcore can be lost on a farm by stoning up muddy gateways, but the timber, particularly damaged or rotten wood, is another matter. During the renovations, as November 5 was approaching, I thought we should raise some money for our general charity fund by having a firework display, and use huge quantities of old timber for a centrepiece bonfire. At one of our impromptu Round Table meetings, around our lunchtime table, I put the idea to my volunteer workforce. Who all thought it made great sense.

One of our number worked in the experimental division of Standard Fireworks in Huddersfield and assured us he would be able to supply us with enough fireworks for a good half hour display. The upside was that they would be free. The downside was that they would all arrive in their brown cardboard cases with no coloured wrappers or names and with no instructions.

We advertised, organised parking, secured caterers and installed the Portaloos. The bonfire was huge. All the local farmers asked if they could bring burnable additions, so we gathered a massive pile of debris. Naturally, we sited it well away from the firework area. We appointed our friend who worked for Standard's as pyrotechnics manager and under his instruction we positioned a fifty-foot bale trailer as a platform for the tamer fireworks. On each side of our stage we dug in three-inch, four-inch and six-inch mortar tubes. We attached eight rocket launchers to each end of the trailer and placed large buckets of sand on the trailer to stick fireworks into. Six wooden stakes were hammered into the ground for Catharine wheels, and we positioned hammers in places where we could find them in the dark. Seven of us were tasked with lighting the touch papers and we had metal bins placed behind each of our positions to keep the fireworks safe.

What we didn't take into consideration was the sheer amount of fireworks we would be getting. They came in a Luton van, full to standing height, and were an absolute jumble. We had a quick rethink and repositioned the van with the cab facing the trailer, so that the fireworks in the open back were not exposed to sparks. Then we seconded another seven Tablers to keep replenishing our bins from the van during the display.

There was little or no light pollution on the top of Cop Hill, and as darkness descended, it brought with it a crystal clear, cold starry night, perfect for the crowds that arrived to see the event. The detonation team worked frantically in Armageddon-type conditions, with ground-shaking thumps from the big mortars, followed by the shriek of the big rockets, bangers going off like machine guns and all of us under a constant stream of sparks and blinding flashes.

The display lasted just over an hour and as it drew to a conclusion, all that was left in the van was a really strange looking firework. It was as wide as a standard dustbin, but about half the height and it had a flat top with a short fuse sticking out of the middle. We all thought it was a good one to end with, because if it was spectacular, then that was a good finale, but if on the other hand it were a bit of a dud then it wouldn't have spoilt what would be the final, frantic spectacular that would precede it.

It took two of us to pick it up and place it in the centre of the trailer, ready for that ultimate moment. It fell on my shoulders to light it, just as the last mortar whoomped its way into the stratosphere. The whole team had gathered a little closer as I lit the fuse. Well, it wasn't a fuse, because the firework exploded, throwing out an almighty black cloud immediately after I touched it with my taper. Everybody within thirty feet was now black from the soot that had shot out of it. When we opened our eyes, we all looked like TV's Black and White Minstrels. Knowing what was about to happen, our Standard Fireworks chap had retreated to a safe distance, and seeing the effect, he rushed forward from the crowd to take pictures of us. He could hardly hold the camera still because he was laughing so much.

We raised a considerable amount of money for charity with next to no outlay and had tremendous fun doing it. Would you be allowed to do it like that today? I think not. Insurance, health and safety, the legal profession have a lot to answer for. They seem to have drained life of so much fun.

Fund raising can be an unpredictable affair, but sometimes somebody comes up with a ludicrous idea that turns out to be not only an immediate hit, but would start a trend that went around the world. One day, one of the Tablers said, 'how about if we sell numbered plastic ducks and race them down the Wansbeck?' The resulting event was an immediate hit, and after four years, the duck race was so popular we had to use a forty-foot tipping trailer to launch the thousands of ducks that had been sold.

We encountered several problems with this scheme, the first being where on earth we could buy thousands of plastic ducks. The first year we sourced a sufficient number quite easily, but as this event got more and more popular we started to struggle, and by the third year, when we reckoned we had exhausted the world's stock of ducks, we had to stop selling tickets long before the event. The fourth year, we ordered a container load direct from the makers in China, with a view to hiring them out to other clubs for fund raising. The next problem with the event's popularity was that the more ducks we launched into the river, the more we had to fish out at the finish, and of course the riverbanks were littered with them the whole length of the half mile course. Then there was the storage of our ducks. Luckily one of our members had a large unused stone barn that he was hoping to get planning permission for, to convert into a trendy property. The problem with this was its location, as all the ducks had to be put into old fertiliser bags and carried 500 yards across a ploughed field. Then there was the numbering of each duck before each race. When we had hundreds it was fun, but when we had thousands it got a bit tedious.

In that fourth year, 1984, we expected both the punters who had sponsored a duck to come and watch the race as usual, and a lot of spectators too. To keep them entertained for a full afternoon we borrowed a field from a local farmer and organised homespun funfair games, such as a coconut shy, throwing balls into a metal bucket, beat the goalie etc. We also got a local butcher to do a hog roast, the local brewery to organise a beer tent and, to complete the day, asked a local pop band to perform so everybody could jig their way into the evening.

About a month before race day the band suddenly became internationally famous. It was *Black Lace*, and their song *Agadoo* was hurtling up the charts. They had agreed to perform for free, and all credit to them, they honoured their commitment. People arrived in their thousands for a free concert by a chart-topping band. We put Black Lace on the back of a wagon and paraded them up and down the crowded streets of Holmfirth. The whole town was ringing and had a Brazilian style carnival feel. Men women and children were dancing in the street and singing *Agadoo-do-do, push pineapples up a tree*. Our conventional little collecting tins were not up to the job so, wearing our Round Table fluorescent tabards, we went back out into the throng, armed with buckets. Every time they were filled, we tipped the contents into bins and went straight back out again.

Midway through the afternoon the duck race did take place, but it was almost a sideshow. We raised thousands that day, and year on year the duck race became more and more popular. Other clubs took up the idea and Pendragon Round Table was able to hire out the thousands and thousands of plastic yellow ducks it owned. As in all things there is of course a downside, and ours was that somebody has to get in the river and collect up all the stray ducks that have tried to escape during the race. On that particular weekend not only did we have wayward ducks to collect, but four dustbins full of small change to count, but this was almost a pleasure rather than a chore, considering what we had achieved.

Kelly's Heroes and feeding experiments

As the industrial unrest of the late 1970s waned, my feeding of Cadbury's Creme Eggs and food factory by-products to the cows at Manor Farm had come to the notice of Kenneth Wilson Ltd, a large Yorkshire-based agricultural feed merchant. News of this sort travelled around the agricultural community like wildfire. Comments such as, 'what's that mad Geordie up to now?' or, 'why's he feeding chocolate to cows? He'll kill them if he's not careful!' ran round the county. It doesn't take long to get a reputation.

In the summer of 1978 Wilson became interested in starting an alternative feed division, but using regularly produced by-products rather than one-off catastrophes caused by missing ingredients, machinery breakdowns and the like. One of the products it was particularly interested in was wet-pressed sugar beet pulp. Although the industry was well used to dry sugar beet pulp, the cost of drying was and still is enormous compared to the extra cost of carrying it. Therefore, if it could be used wet, it could be marketed a lot cheaper.

There was a distinct lack of knowledge about how much a cow could eat a day, so they wanted some data about how to work out feeding regimes, etc. I was particularly suitable for collecting this data, as I milked in a shippon and bucket-fed my cows individually when they were chained during milking. The company also felt it could capitalise on my experience of introducing new feeds to cows and dealing with whatever problems ensued. Clearly, I could not be paid to provide this information, or claims of bribery could be made, rendering any results I achieved commercially worthless. Instead, Wilson provided me with the infrastructure to handle the forty-foot tipper wagons and a hard pad for them to tip it onto.

The day the farm track was replaced with a proper metalled road arrived none too soon. The day before, the milk marketing board tanker had sank up to its axles in our rubble lane as he was leaving, and, with 30 tons of milk sloshing around inside the stainless steel tank, the combined pulling power of all the surrounding farms' tractors could not budge it. A second tanker was summoned and the milk was transferred to it, then the local tractor tug-of-war team inched the wagon out of its self-made grave.

The hole it left behind in our drive was enormous, but the next day the cavalry arrived in the shape of Patrick O'Grady and his merry entourage. All we needed was the music from *Kelly's Heroes* playing as they rose over the brow of the hill. First in the parade was an enormous bulldozer, followed by a road roller, a tarmac laying machine, a road plane and finally a series of huge tipper wagons. The vehicles had about twenty brawny individuals hanging on to their sides, standing wherever they could find purchase. These weren't your average drive-laying types. They were hard boys, tar stained, commercial motorway contractors and completely out of place in this rural idyll (well, it was my rural idyll).

Where Kenneth Wilson had found them, and at what cost, I have no idea, except we were only about four miles away as the crow flies from the M62, where some serious remedial work was being done. It's amazing what a big roll of the folding stuff will get you.

The only words exchanged were, 'mornin' gaffer. This'll be Upper House Farm so be it.' Pure Falls Road.

'Hello, yes this is Upper House. Shall I show you what...'

'I've a paper on what's to be doin', he said. 'Me and the boys 'll be out yer hair in time for the bar opening.'

And that was it. After the opening words, there was no more conversation with Patrick, or any pleasantries with the rest of his crew. It was as if they were deaf and dumb, but in four-and-a-half hours they dug out and extended the old road, laid a foot of graded compacted hardcore, then laid about four inches of that really hard, rubberised motorway tarmac, leaving me with 250 yards of road of a better quality than all the lanes in the surrounding area, a turning area for the large articulated trailers and a large yard to tip the pulp onto. And at the end of their toil, they disappeared like snow off a dyke.

The next day, 25-tonne artics from the sugar beet factory in York rolled into the yard, and spewed out their load of grey sludge onto the pristine tarmac. To get the last dregs out of the body of the trailer, they had to tip it to a vertical position, making it almost 50 feet high. Unfortunately, our overhead electric cables were only 30 feet off the ground. On the first day, a tipping wagon snagged the cables and lifted them high into the air. We all panicked and called out the Yorkshire Electricity Board to solve the problem. When the engineer arrived he looked at the situation, went to his van, got out a telescopic pole and a thick rubber mat and told us to stand back. He placed the rubber mat on the ground and stood on it while he unhooked the cable with his fully extended pole. He explained to us that the cable between the poles on Cop Hill was so slack that lifting this one up had done no damage. So, the next time it happened we used an extremely long sapling as a substitute for the engineer's pole and an old tyre for insulation. It did the trick and it didn't waste four hours of our time waiting for the Electricity Board.

Feeding the pulp to my girls was really amusing. I had been feeding the concentrated part of their ration, commonly known as cake, at milking time. It is very dry and is made by compressing grain, pulses and soya into pellets. We call it cake because they love it. If you allow them to eat it *ad lib* they will gorge themselves until they are ill, sometimes going as far as death. A high yielding cow may well eat 10kg or more of cake a day but to achieve the same level of nutrition with the wet pulp the ration would weigh 78kg. If fed twice a day that would mean 39kg each milking (about half a wheelbarrow-full). It takes about five minutes to milk a cow while she is in her stall, and getting her to eat such a large volume of pulp in such a short space of time seemed like quite an ask. As it turned out, I needn't have worried, because the cows all learned to suck it up like they were drinking it, and by the end of the experiment were able to take in twice that amount, and more, if they were given the chance.

So what you may ask was amusing? It was the noise of fifteen cows sucking and swallowing. If you could imagine a room full of children having finished their strawberry milk shakes, searching for the last dregs with their straws but only ten times louder. The quantity for each cow at each feed amounted to three or four buckets full, and when the buckets were empty and the cows were impatient for a refill they would hook them with their noses and throw them into the alleyway. The buckets were galvanised metal and, as you can imagine, the noise was deafening as fifteen of them crashed onto the concrete floor.

Providing this service for Kenneth Wilson's had several veterinary implications. I did say that if they ate too much, they could kill themselves, and I'll explain why. Over the months we built up the quantity of pulp we fed to the cows and they started to reach saturation point nutritionally, but if the cows were allowed to eat too much it might cause bloat in the rumen, a large fermenting organ which is part of a cow's digestive system, located just before the four stomachs. Bloat happens when an excessive amount of gas given off by the digestive system jams the valve at the top of the rumen shut. This causes the rumen to swell to such a degree that it presses on the heart and stops it dead. Also, the research department wanted to see what would happen if we exceeded the recommended calorific intake, or the metabolisable energy (ME). These possibilities sent me scuttling to the vets in Huddersfield to get stocked up on the various drugs, chemical treatments and surgical instruments and to get some medical advice, and it was on one of these trips that I made a fascinating discovery, not related in any way to the feeding experiments.

It was on a sunny afternoon that I had arranged to go to the vet's. I had spent all morning mucking out the cows' winter quarters, and in my normal way of over estimating what I can achieve in a day, I had not left enough time to get washed or changed. In the sultry afternoon heat and still in my working clothes, I must have looked like Claude Greengrass off ITV's *Heartbeat* as I made my way down Byram Street to the vets. The eau de countryman fragrance I was giving off must been overpowering.

Part way down the street I passed an art gallery, and as I did so, glanced in the window. Sometimes it takes several moments for my grey matter to work, but I realised I was looking at an oil painting of my house. It was big – about four feet wide and three feet high and my immediate thoughts were that the artist must have been trespassing when he painted it. Then I thought, 'wow it's bloody good!' It captured the wildness of the steading's setting and the clean lines of the modernisation, integrating a late eighteenth century shippon with the stark whiteness of the modern concrete barn.

I was in through the door like I was being chased by a Doberman Pinscher. 'How much for the picture in the window?' I asked the man in the gallery, whom I assumed was the owner.

'That picture, as you call it, sir,' he said in an extremely snooty voice, 'is a very rare work of art. It's an oil painting by Ashley Jackson. He usually only works in water colours.'

'I've never heard of him,' I said. 'And that does not answer my question.'

'250 guineas.'

In today's money, that would be about £1000. 'Well,' I said, after getting my breath back, 'you'd better give me his telephone number, because I want to know what right he had to trespass on my land when he was painting it.'

'Ahh...'

I looked him in the eye. 'I'm sure we can come to some agreement on the price which would be amenable to us both,' I said. The heat of the moment was passing and I became increasingly uncomfortable of how I must look and smell, and as I didn't know who Ashley Jackson was, I was clearly a philistine in the eyes of the gallery owner. Jackson, so I was told, was internationally famous. Barclays Bank even had a collection of his small watercolours as cheques, writing the amounts and the payee's name across them. As long as the commercial value of the painting was greater than the value of the cheque, it worked. The banks never defaced the paintings by using their normal ink stamp and consequently they were never properly cashed, but the bank honoured the value. That's a neat trick, if you're a good enough artist. Then the gallery owner suggested I commission someone else to paint a picture of the farm. However, I gathered my wits.

'My name is not Duke Philip of Cop Hill,' I told him. 'It's Farmer Philip and there's no way I'm paying that sort of money.'

Then he changed his attitude, saying that one of the less famous artists who supplied the gallery would paint it virtually for the cost of his paints and there would be no obligation to buy. What did I have to lose? He gave me the names and addresses of a few artists and I went on my way to the vet's. A few weeks later I contacted one of them and arranged for him to come over. He charged me £30 for the materials and offered me a choice of four different paintings. But it occurred to me, thirty or more years on, that as investments go I had perhaps missed a punt in not buying one of Jackson's paintings, so I recently looked up the current prices for his work, and although he is clearly very famous, the equivalent value of his current paintings didn't look a lot different from when I'd first seen his work. I suppose I would have had to wait until he was dead before prices rose significantly. That's a bit sad for him – imagine all those callous art aficionados waiting for his demise so they could cash in on their investment.

My continual search for dairy equipment took me on a day trip to Robin Hood's Bay on the east coast. Of course, you can't leave the coast without buying a healthy portion of seafood. I bought two extra-large live crabs, which I transported to the farm in a bucket of seawater, covered in cling film and jammed into the corner of the boot.

I arrived back just in time to start milking, but with not enough time to retrieve our tea from the car. At the end of milking I sent the children upstairs to jump in the bath together: a custom that had started in the dairy trough and one which they all enjoyed. I went to retrieve the mystery tea from the bucket in the car and, as I hadn't told the children I'd bought the crabs, the temptation to play a trick on them was just too much. I crept upstairs with a crab in each hand, fingers firmly spanning their backs to avoid getting nipped, because they were wriggling their legs and pincers to free themselves from my grip. Dropping to my hands and knees I silently eased the bathroom door open, just wide enough for me to squeeze my way in and keep below the children's line of sight. The bath was in the corner of the room, and at one end there was a large flat area to sit on. It was an ideal spot to place the lively chums on and it only took seconds for the fun-filled three in the bath to notice them. There was a triple-voiced scream and together they shot up and exited the far end of the bath. Unfortunately for me, they took most of the bath water with them. Downstairs was flooded. Fortunately, the ceiling had open beams, but the carpets were soaked and the lights fused. Was the hassle worth it? You bet, and fresh crab on hot buttered toast went down a storm as well.

A turkey shoot, and decision time

Merrydale is a small valley above Upper House Farm, and where the Merry kissed the lonnen (the lane) was a collection of stone buildings and houses in various stages of disrepair. In one of these lived a colony of, not quite hippies but extreme bohemians, real hand knitted yogurt types. They were intellectual sleggers, (scroungers) drawing the dole, smoking spliffs and trying to live in a self-sufficient way, but with absolutely no knowledge of what they were doing. The place was a mess. The gates were left half open and hanging on by one hinge, the vegetable patches were partially cultivated then left to go to weed patches and there were scrawny chickens and ducks wandering in and out of the houses and onto the road.

One evening while I was milking, a human scarecrow wandered into the byre. He had greasy, mousey brown hair down to his waist, tartan bell-bottom trousers that were so dirty they looked plain brown, and a long, loose knitted gansy. His name, he said, was Rupert. He was extremely well spoken but his speech was impeded by an overgrown moustache and beard and by the butt end of a roll up, superglued to his lower lip. He had acquired some turkey chicks that he wanted to fatten for the Christmas trade and he wanted a quick lesson in husbandry for rearing them. After an hour long lecture on heat lamps, turkey rearing mash, constant access to water and the like, I got a high five, a 'cool,' and then off he went in a slovenly, lollopy gait, bending at the knee with each step and over-swinging his arms. He moved like that character on the Kia Ora advert that was showing on TV and in the cinemas at the time... boogie, boogie, boogie.

A few days later, instead of going straight down to the town as I usually did, I went in the opposite direction, a route that took me round the head of the dale and down the other side of the valley and past Merrydale. I was curious to see how the chicks had fared. As I rounded the corner I was greeted by perhaps two hundred grey balls of fluff running all over the road. I pulled up and went in search of Rupert. He was fast asleep in a deck chair, mouth hanging open and what looked like the same butt end stuck to his lip. I told him about his flock of chicks escaping and he explained he was going to raise them free range and just let them roam. I warned him of foxes and the need to house them at night. 'Cool,' was his only reply. I left him to it.

Upper House was reasonably isolated; the few neighbours we had were Ronald and his brother, Edwin, the Rose and Crown pub, and Don, who had helped me move my baker's table. Don's house bordered the farthest corner of our land and he lived there with his young family.

Most people regard rabbits as cute furry creatures that eat grass. They live in their thousands in the woods surrounding the farm and are a real pest, so during that summer, when Don asked if we minded him snaring a few rabbits for the pot, I readily agreed. A land owner or tenant has the right to hunt and consume all game and vermin that lives or strays on his or her land, so I agreed to take half of Don's catch, which was the normal rate for that sort of thing. He was taking his children along for the walk while he set the traps and asked if Sharron wanted to come too. She liked the idea and I was happy for her to go, so off they all went and had a pleasant morning as Don secured tomorrow's dinner for us all.

During milking that afternoon my senior assistant was Tracey-Emma. She was only four but she had no fear of the cows. They were curious but gentle with her, although when you have a little one around you have to err on the side of caution. Whenever the cows were loose and moving freely around the shippon she would go outside and stand on the dairy steps out of harm's way and then, when all the girls were fastened back up, I would call her back in. This particular time she didn't return and I went outside to look for her. She was nowhere to be seen. Thoughts of Sharron sitting under the bull flooded back to me, even though at this point we did not have a bull. I ran out into the nearest field, calling her name, when my voice was drowned out by the drone of a microlight aircraft flying overhead. The pilot shouted down to me that a little child was in the next field behind one of the drystone walls that border all the fields on the farm. I was waving my thanks to him when a low flying Tornado Jet fighter screamed over the brow of Cop Hill brow and passed within what seemed like inches of the microlight. The tailwind blast of the jet sent the fragile craft into a fearful spin and I thought he was going to crash, but he managed to take control once more amid a stream of colourful expletives emanated above the disappearing roar and he put-putted off into the distance.

I found Tracey-Emma where the microlight pilot said she was. 'Where on earth did you think you were going?' I demanded.

'Hunting rabbits, like Sharron!' She said.

'Sharron's' playing at home now,' I said. She had returned late morning just in time for her lunch.

'No she isn't,' Tracey-Emma replied. 'She's in the woods.'

That's curious, I thought, and I decided we would go and take a look. As we approached the trees I thought I could hear singing, and sure enough, as we followed the sound we found Sharron sitting down in the middle of a warren, singing *Bright Eyes*, the song from *Watership Down*. I asked her what she was doing and she said, 'Uncle Don said, if I sing that song down the holes the rabbits will come out and play with me.' So there she was, singing her little heart out hoping the bunnies would come out to play with her. She was obviously blissfully innocent of the darker reason for summoning them.

As that summer passed into autumn, I started to get curious again about the Turkey rearing at Merrydale, so, one day I drove round and much to my surprise there was a forty-strong rafter (perhaps a better group name would be a 'gobble') of well-feathered, healthy looking, immature turkeys. Further on I found Rupert, who was trying to repair one of the many drystone walls that were continually collapsing. I told him how well his flock looked and he grimaced, saying the fox had got the others on the nights he had forgotten to put them away. I told him that he would have to clip their wings or keep them shut in a building, otherwise they would fly away.

'No need, man,' he said. 'I feed them too well. They're too fat to get off the ground.'

This utterance reminded me of those famous last words of King Harold at the battle of Hastings. 'Watch where you're waving that bloody arrow, lad,' he said, 'or you'll have some bugger's eye out.'

Sure enough, three weeks before Christmas, forty turkeys took to their wings and flew into our neighbour's woods. In a situation like this the ownership of the turkeys comes under debate. You have a duty to fence against your own stock but remember what I said about game and vermin straying onto someone's land. If, for instance hand-reared pheasants stray onto neighbouring property, the birds are considered in law to become the property of the neighbour.

Rupert and his cronies tried all ways to entice the birds back, with absolutely no success. They had gone feral within days of taking flight and unfortunately for him he had presold thirty of them for Christmas. A delegation from their encampment arrived at my door, asking if I had any suggestions. The woodland was not mine to speak for, but I suggested a turkey shoot, like they have in America. I was sure a day's shooting for four or five guns would be well received, if the guns got to keep a couple of turkeys each. Of course the landowner's permission would have to be gained before anything like that could commence. I told the group I would arrange for the guns and they could go and get permission for the shoot. That set them all off grumbling and moaning about having to put some effort into sorting their mess out.

A few days later I met Don with two of his friends at one end of the ancient nine-acre woodland that bordered each side of the Merry. We were all good shots, well used to bagging pheasant and grouse. By comparison, turkeys are enormous and the consensus of opinion was, the turkeys being much bigger targets it would only take a couple of hours to bag them. The surprise was that after three solid days of shooting, we had only managed to bag eleven turkeys. The rafter was incredibly wary when we started but by the end of three days, if you so much as breathed a bit heavily, they were gone. They achieved their Houdini-like escape by staying in the cover of the woodland, then, using their weight to blather their way through the lower branches, they disappeared in seconds. Over the next few years we occasionally had a pop at one or two that ventured out into our fields. We certainly never managed to shoot them all, but they gradually disappeared, so I can only assume that foxes, the climate or old age finished them off. Certainly they never started breeding, more's the pity.

Our second winter at Upper House was really severe and I spent many hours clearing snow with my old John Deere tractor and rear loader. It was imperative I cleared the roads so that my bottle milk buyer could collect his daily wares, and also to enable the Milk Marketing Board tanker to collect the balance of my production. I had found an in-crate bottling machine in the corner of the dairy hidden under a pile of rusty milk churns and had renovated it and put it to good use, bottling 50 gallons of milk a day for a local milkman, so I was gradually building up the income from the farm. However, winter in the Pennines brings its own problems and forced me to face up to some hard realities.

As soon as the snow started to fall I would be out there on the tractor, ploughing it away, keeping us open for business. One evening I spent five straight hours on the cab-less tractor, in sub-zero temperatures. I had not realised how much the cold had penetrated and when I came to climb out, I found I was frozen to the seat. I took off my heavy wool greatcoat and managed to escape. Clumsily I tottered my way towards the warmth of the Aga-heated kitchen when I slipped and fell onto a wheelbarrow full of the night's semi-frozen detritus from the shippon. It tipped over, throwing me onto my knackered arm and jamming me up against the large wooden sliding door into the byre. I couldn't move and almost instantly my damp clothes froze to all the icy surfaces. I was stuck fast and the searing cold was penetrating deep into my core.

For how long I lay there before Sharron, then seven years old found me I have no idea, but long enough for me to realise I had been a fool to think I was going to manage this lifestyle long-term. The Round Tablers would get fed up with helping me and relying on my family was not fair. I had to find another way of making a living and little by little, over the next months while I was doing mundane tasks out on the farm, my mind would be in overdrive, thinking up ideas.

At that time the dairy industry was controlled by the government-run Milk Marketing Board, but under pressure from Europe, the government wanted to end the Board's monopoly. This made the dairy farming community very nervous, because it meant that the large dairy companies would control the price they received for their milk. At all of the farms I had worked on, we had done something with the milk, be it separate milk for cream, put raw milk into bottles, or pasteurise it and put it into cartons, and it invariably doubled the farm income. The ever-growing problem was that you only had one product to sell. Customers, be they wholesale bottle milk buyers, retailers, or householders, expected a full range of dairy products. The big dairies realised this and progressively refused to sell to small wholesalers, so as to limit their potential in the market place. Around the same time the food hygiene revolution had begun and individual farms had neither the technical knowledge nor the finances for the laboratory needed to comply with new rules and regulations.

Then I had a eureka moment. Supposing there was an organisation that could co-ordinate groups of dairy farmers to produce the simpler products, in packaging with their names on, but sold under one name? That organisation would produce whatever products were missing from the range, provide laboratory facilities, a sales team and technical knowledge. So, the Merrydale Dairy idea was born. Even today, I think it's a good idea.

I needed more time to develop this project, so to cut down on the time I spent on the farm, I decided I needed a dog to gather my cows for me. My search took me to Saddleworth Moor, where I bought Sweep, a long-haired Border Collie who was being retired from the hill at the ripe old age of seven. When the farmer demonstrated him, it was like watching the old TV programme, *One Man and his Dog*. Sweep responded to his master's slightest cue and I thought he was exactly what I needed. As I had never worked with a trained dog before, the farmer asked if I could whistle. I gave what I thought was a reasonable account of myself by way of demonstration.

'Aye, ah thowt.' He said. He clearly thought my effort was abysmal. 'Yi'll be needin' one o' these.' He put his hand into his pocket and pulled out a small piece of stainless steel, blew off the fluff and bits of hay stuck to it, and handed it to me.

It was a sheepdog whistle. It was made from a metal disc, two inches in diameter, that was bent in half with a small hole in one of the faces. He quickly explained to me how to place it on my tongue, curve it against the roof of my mouth and blow across the hole. It sounded simple, so Sweep and I, along with my newly acquired whistle got into the Land Rover and started to make our way home. I popped the whistle into my mouth, thinking I could practice as I was driving, but apart from spitting all over the inside of the windscreen, the only noise I managed to make was a wet, farty one. The whistle shot out of my mouth several times and eventually ended up on the floor of the Land Rover. This was going to be tricky. It took me a month to perfect the art. In the process the whistle gave me a very sore tongue and all the saliva I had dribbled while practising gave me a chapped chin.

When I got Sweep home he was in dire need of a good brushing, but he was a very macho dog and no human had ever made him look poncy before. He was too old to change the habit of a lifetime and when I approached him with a brush, he backed away and growled menacingly. I carried on and he bit me four times before I gave up. From then on, he remained a scruff-ball and we tolerated one another – just.

As for working, he was mustard, just so long as I kept the commands verbal or the whistle blasts very quiet and low key. If I were too demonstrative with my commands, which often happens in the excitement of any gather, he thought he was back on the moors and he would be over the wall like greased lightning, streaking out to the other side of the valley. The louder I shouted, the further he thought I wanted him to go. Arriving at the location he thought the command had sent him, which was usually a neighbour's field, he would gather the field of stock to the bottom right hand corner, then jump over to the next wall until all the stock, be they sheep, cattle, geese or hens, were all neatly cornered. When he felt his work was completed, he would make his way home. Fortunately, my neighbours never complained, as he did no real harm. However, it did encourage me to control the level of my voice. Nowadays, I try to tell friends and family that Sweep taught me a valuable lesson; to be a quiet, discreet Geordie when situations require. Normally, those who know me fall about laughing and look at me as if I am quite mad.

Around this time we were given a Border collie puppy, which my children loved, as he was always keen to play with them. That is, until I managed to squish him with my tractor. The children were distraught because my other dogs were grumpy and not exactly children friendly. So I asked the local branch of the RSPCA and told them that the next time they had a Border collie puppy that needed a home, I was their man. I didn't have to wait long before a phone call summoned me to the kennels. There I was given a creature that only just covered the palm of my hand. 'If you can keep it alive,' they told me, 'he's yours. He's very young and we think he's a collie.'

I had gone to collect the puppy on my own because, then, if it wasn't suitable I could be hard and say no. So, was this adorable, blue eyed, ball of blue grey fluff going to be suitable? No! So what did Hard Lad Dixon say? I refuse to answer that question on the grounds that it may incriminate me, but I called him Plod and I can tell you that I have never named a dog so aptly. Over the next few weeks he grew like Topsy. Within a week he was Jack Russell size, within a month he was as big as a cocker spaniel and at three months he grew bigger than the biggest Border I have ever seen. He was a sheepdog, but an Old English one and he was the thickest canine creature I have ever come across.

A five bar gate led into our yard. Sometimes it was open and sometimes it was closed. When it was closed, Plod would run headlong into it and then stagger back a few paces and have another go, banging his head again. Just like the Dulux dog, he had hair all over his face, so to assist his vision we cut it off. Did it make the slightest bit of difference? No, but at least we could see the look of confusion on his face. Perhaps you might excuse his ineptitude on the fact that an Old English sheepdog's eyes are set on each side of its head, giving no direct forward vision. But then why, when he staggered back from colliding with the gate, and often tottering sideways, did he not notice this ten feet wide, four feet high obstruction? And when I emptied the hot ash out of the Aga onto the midden on a cold dark morning, leaving several red glowing coals in amongst the ash, Plod would bimble over, sniff at the debris, then put a paw onto the hot coals, yelp, look in disgust at his paw then put it back on the ash and singe it again.

Plod was a big cuddly toy, and within months of us getting him, he dwarfed Gavin, who was then three years old. The two of them were as thick as thieves and spent most of the day together, getting into all sorts of mischief. One day, I heard a yell from Gavin. I rushed over to find Gavin in floods of tears. Plod had bitten his nose and had drawn blood either side of the bridge. I was furious and turned to shout at Plod, who by then was cringing and whimpering in the corner. As I got closer, I noticed Plod was bleeding as well. His cheek flap was almost split, with little teeth marks in it. I went back to Gavin to find he had a mouth full of Plod's hair. The little monkey had bitten Plod first! To this day Gavin has the pale outline of a scar on his nose.

www.velocebooks.com / www.veloce.co.uk
All current books • New book news • Special offers • Gift vouchers

Get fresh daily

On September 8 1981 the venture that had been hatching in my head for some time became a reality when I registered Merry Dale Dairies as the name of my limited company. As its slogan I chose "Get Fresh Daily." It was a great accomplishment and one that I took huge pride in, but, as we all know, pride comes before a fall.

As a young, inexperienced businessman I could not anticipate the depths my competitors would sink to in order to gain the upper hand. What happened is true, but why and how Merry Dale was destroyed is just unsubstantiated theory, as I have no proof of what happened outside of my own experience, or what part any of my alleged opponents played. To avoid libel, I have changed a few names.

To get to the point of incorporation I had written a five-year business plan, and an explanation of what I wanted to achieve. I got my brother, Noel, involved at an early stage. As he had a science background, he became science director. This, coupled with my dairy business experience and agricultural qualifications, was the backbone of our theoretical and practical skills, but we needed money. Between us we had about £20,000; we needed about £250,000. Normally, banks lend up to an amount equal to what the borrower has to invest, therefore, theoretically our funds were £40,000. No matter how bad you are at maths, it's obvious that we were £210,000 short, but with the youthful blind belief that it was a really good idea, we soldiered on, looking for a home for our new venture.

We found it in Consett, Co Durham, which is 120 miles away from Upper House. Maggie Thatcher's thirst for curbing trades union control resulted in the closure of Consett's steel works, and dissolution of the mines supporting it. This resulted in an unemployment level of 38 per cent, so the town was hungry for new businesses and employers. The European Union had given the UK government millions to spend on installing a new infrastructure, but the new roads led to nowhere: bare fields with drain inlets, high voltage electrical points and mains water, all just waiting for new industries to materialise.

I had an appointment with the business development officer of Consett's town council at 10 o'clock one morning. His office was located in the large, prestigious Victorian town hall, a building that reflected 150 years of the town's once proud steel-making heritage. As I entered through the large carved oak doors, the silence was deafening, and when the door slammed behind me, the noise echoing around the cavernous interior was like a cannon going off. The entrance hall's ornate decor, with heavily embossed tiles, wrought iron bannisters and sweeping, polished wood staircase wouldn't have been out of place on the set of Gone with the Wind. Once it had given an impression of grandeur, but now it just felt empty and sad.

Nevertheless, the smell of beeswax and disinfectant in the air told me that Consett's sense of civic pride still survived. The reception desk, surrounded by tired, faded posters was unmanned, so I had no idea how to make contact with anyone. Then I heard from high above me the sound of a door banging shut and a wet fish of a man slapped his flatfooted way down the stairs towards me. He managed to make the brown suit and red tie he wore look grey. Here was a defeated man. 'Mr Dixon,' he said. Despite his appearance, his handshake was strong and dry. 'My name is Brown. I'm the business development officer. Would you care to follow me?'

We made our way up to a large room at the front of the building and in contrast to the rest of the building, it had been modernised, with a false ceiling, fluorescent lights and word processor workstations, partitioned by moveable screens. All that was missing from this state of the art office suite were people. Mr Brown seated himself at a small conference table by the window and invited me to sit beside him. I drew a deep breath, and began my sales pitch for Merry Dale Dairies. After a few moments he drew a pad towards him and started taking notes. His resigned face took on an expression of curiosity and as I rattled on, his face changed once more. He was catching my enthusiasm. At the end of my spiel I showed him my cash flow forecasts for the next five years and as he looked at them he stood up and with a raised voice said, 'Let me get this straight: year 1, build a dairy, install processing equipment, a bottling line, a carton line and a laboratory; year 2, consolidate sales and reputation, start meeting local farmers about processing products and selling the Merry Dale idea; year 3, get five farms on board, producing cream and yogurts, grow the dairy sales and throughput; year 4, bring another ten farms on board, making soft cheeses, ice cream and butter and year 5, engage more farmer partners, producing multi-layer desserts, and create modest dairy growth.

'You would be employing 15 people straight away and by the end of 5 years, 60-plus people on site and creating employment of a similar size in the countryside around, making it 120 jobs. Have I got it right?'

'Yes,' I said. 'I think that's a fair summary in the present market with the MMB still in place. But if the government scraps the monopoly of the milk marketing board, the growth could be exponential, and remember, about half way through year 3 we get to a stage where it will be almost self-financing, because the farmer partners will be buying their own capital equipment.'

He was now hopping from foot to foot. Sweat was on his brow and this shell of a man had transformed into an excited schoolboy, his eyes shining bright. I was very reluctant to rain on his parade, but I pointed out how little capital Noel and I could raise.

He brushed that to one side with a sweep of his hand and said if I had time could he take me to Durham City to meet the county development officer. I told him I had all the time it took, so off we went to Durham.

By the end of that afternoon, having had to say very little, I had a small team of people around a table all talking as if it was all going ahead. Phrases like, 'this could be massive,' or 'keeps jobs in the countryside,' 'innovative – we've got to make this happen,' or the more down to earth, 'Jesus! At last a new idea!'

I was thrilled with their enthusiasm, but at the end of it they admitted they were political animals, not businessmen. They would appoint a business consultant on my behalf, to look at my figures and provide them with an independent feasibility study.

They appointed a company of business consultants in Newcastle. I took all my plans and figures along to the MD and sold him the Merry Dale project. Despite him being the MD, he was a one-man band and although he had a prestigious address and office, I did not warm to his greasy, slime-ball smoothness. He was a slightly hunched six-footer and drainpipe thin. He wore a dark blue suit with wide pinstripes and had slicked back hair and a pencil style moustache. He was Dick Dastardly in a business suit. Did I trust him? I didn't even consider it. I didn't like him, but I had little choice. He constructed a favourable report using my figures, and presented them in a different format to each of the various interested parties, claimed them as independently verified (although I don't believe he cross-checked anything), and collected a handsome fee.

An endless round of meetings then took place with ever increasing numbers of participants from various local, county and national government departments seeming to need to jump onto the bandwagon. I had thought my biggest obstacle was money, but that proved to be the easy part. Grants were available from the town council, the county council, the Ministry of Agriculture and the EEC (Consett had been declared a least favoured area) for all sorts of aspects of the project. There were capital asset grants, quality assurance grants, feasibility grants, mechanisation grants and job creation grants. All four agencies worked together and their ability to say, 'yes, we can do this' was astounding. Where funds fell short and we needed to borrow, local authority backed guarantees were available to placate the banks.

A site in the middle of the new Watling Street Industrial Estate was allocated to us. Already in place were roads, streetlights, broad raised pathways and roundabouts, but the main roads went nowhere and the streets leading off the junctions finished about 10 yards into open grassed areas. It looked like a very large Scalextric track, laid out in the middle of 100 acres. It was very surreal.

Within months, the steel skeleton of our building, the first to be built on the estate was standing on the skyline like a sentinel for the future of the area. The local steel fabricators and builders must have been waiting on starting blocks desperate for the first enterprises to arrive, because the speed at which the building went up would have sent TV house building specialist Kevin McCloud into paroxysms of admiration. The shape and size of the dairy was predetermined by the "off the shelf" factories that were on offer by the County Council. Even the design of the security fence surrounding the site was out of our control.

When Merry Dale was in its early stages, I had to spend an increasing amount of time away from home, so I employed a young farm worker, Chris, to do the physical stuff while Carol managed the rest. Even with the 119-mile commute to Consett I got home most evenings, usually in time to bath the children and put them to bed. Twice a month I relaxed with my fellow Round Tablers, my friends who had showed me and mine such kindness, although it must be said, I had neither the time nor the money for this indulgence.

In the summer of 1980, Pendragon Round Table had been invited to go canoeing with a local club at Penistone quarry. As was usual with my over-full life, I arrived at the gravel pits too late to receive instruction and all the available canoes were being used. After half an hour, one Round Tabler, Charlie, paddled his way back to the shingle beach from where the flotilla had launched itself into the peat-stained water. Getting him into the kayak, they told me, had been easy, his 18-stone weight adding sufficient force to ensure entry into the narrow throat of the bright yellow stunt canoe. His belly filled the cockpit entirely; in fact it hung over the bulkhead, which made the canoe look like some sort of bizarre prosthesis.

But if getting him in the kayak was easy, getting him out of it was like trying to extract a Zeppelin from a cat's bum. Three of us who were on the shore tugged at him and he yelled in pain. We dragged the kayak, with him in it, on to the shore, but all we managed to do was get very wet. When the rest of the club paddled back, six of us tried to get him out, with three of us pinning the canoe to the ground and the other three pulling at his torso. He didn't budge. He just yelled louder. In my car I had some calving oil, a veterinary version of KY gel. I went and got it and told him to rub it round his belly so that it got between him and the rim of the cockpit. Then we had another go at tugging him out, with no success. The situation was silly enough as it was, but then some bright spark suggested we should leave him in situ and go to the pub with him on the roof rack of Mike's Land Rover. We all voted what a good idea this was, except for Charlie, who cursed us and inferred we were all born out of wedlock. But when fifteen blokes get together to move one immobile person, then no matter how big he is, he goes.

We lifted the canoe and its contents on to our shoulders and carried it towards his new, highly polished black 110 station wagon. It was at this point Mike started having second thoughts about using his new vehicle for this. Unfortunately for him, a lynch mob mentality had taken us over and we lifted the canoe above our heads. This is when our plan came unstuck, because there are several places to hold onto on the top of a fibreglass canoe, but the underside is as smooth as a baby's bottom. The result was that Charlie's prosthetic appendage slipped round and turned upside down, leaving him dangling upside down and squealing like a giant fruit bat. We hadn't thought of this method of extracting him before, but some bright spark suggested we should shake the yellow banana up and down. It worked. The calving oil had spread around with the movement, and with him upside down, his belly was folded over his chest. Gravity took over, and with the noise of the biggest, longest and wettest fart he started to slide out. He hit the ground with his shorts around his ankles. It took every one of us, Charlie included, half an hour to stop laughing.

The Tablers had arranged to eat at a local pub, but I was wet and dirty and had missed out on the water play, so I decided I would go for a swim in the lake and join them later. As the noise of my food-hunting friends' cars disappeared into the distance, a peace settled on the ancient industrial excavation, where Mother Nature was fast reclaiming the lake from the man-made scarring of the countryside. She had made it hauntingly beautiful and the atmosphere was only interrupted by the occasional skylark claiming the clear open skies for himself, leaving the summer warm water for me to gently slide through. It was so delightful that I swam for about an hour.

When I climbed out and dried off, I tried to find the pub where the rest of the crew would be. I followed the instructions I had been given to the letter, or so I thought, driving up and down every lane and track in the area, but I couldn't find the place. I was getting really hungry and I knew there was a roast on the go at home, so I called it a day and made my way back to Upper House, thinking about my dinner. It would be clay cold by now, but after ten minutes in the hot Aga it would be lovely.

By the time I got home, it was nine o'clock and still daylight, the sun having just sunk behind the Cop. It was unheard of for me to return so early on a Table night: normally, it would be nearer to midnight. I parked the Saab and went in, but didn't shout my usual 'hello,' as I knew the children would be fast asleep in bed. The stereo was playing quietly in the sitting room but nobody was there, so I switched it off and, as I made my way to the bottom of the stairs, I heard Carol in our bedroom. When I went into the room she was standing at the foot of the bed, with a strange startled expression on her face. Her clothes were dishevelled, and as I swept her into my arms I asked her if everything was all right. She said she was fine, but then she slid her arms down level with my elbows, locked her hands behind my back and shouted, 'RUN, CHRIS, I WON'T BE ABLE TO HOLD HIM FOR LONG!'

It took a few moments for the penny to drop, but then with a roar of pain that came from my heart, from my trapped broken arm as I tore it from her grip, and from my head as it struggled to compute all this blinding information, I ran out into the hall to see Chris, my nineteen-year-old farm labourer, legging it down the stairs. Half-clothed, he ran out into the yard to his beloved motorbike, but I was too close behind, so he ran past it. I stopped there, looked at this highly polished bike, and with adrenaline coursing through my veins, picked it up (God knows how) and threw it after him, smashing and twisting the fairing. I returned to the house and, as I reached the back door, heard the roar of his bike as it gravel-spun off into the distance.

I have very little recollection of what happened after this point, but I think Carol and I talked, rowed, fought and tried to make up, but in the end she told me that she loved chris and was going to leave me. I was devastated. She had been my first love, and my only love. She explained that after my accident, if I touched her with my deformed arm it made her shudder. I was never at home, she had no life of her own, and she felt trapped in a life not of her choosing. To her, everything in every aspect of her life was the pits. To be fair, reflecting on this over three decades later, it does not seem to me that we had much of a partnership. Because I was so busy trying to succeed, I had overlooked her needs, and probably taken it for granted that she was happy simply providing a home for the children and me. She left it all, children included, and moved out within days. So, there I was, not quite 30 years old, three children to care for, two businesses 200 miles apart to run, and no one to run the farm. To cap it all, Chris, the cheeky bugger, sent me a bill for fixing his fairing. Needless to say, I didn't pay it.

And what do big boys do in a situation like this? They dry their eyes, get off their backsides and go home to Mam. (The Hyacinth Bucket in my mother still hated being called "Mam.") We arrived en masse at New House in Gateshead and Mam did fix it, until we reorganised our lives. The first decision was whether to keep the farm or forge ahead with Merry Dale. As Noel, together with the combined agencies of the North East, had invested a vast amount of time and even bigger sums of money in Merry Dale, I chose, reluctantly, to give up my farm. This was going to take some time to arrange. I had thought my life was busy in the past, but at this stage I was so pulled in all directions that I thought my head would burst at times. But as Mr Hall, back at North Farm, would say to me, 'if a problem has a solution, then it's just another job to do, and in farming there are always more jobs than time, so you just have to learn to prioritise.' So that's what I did: family first; Merry Dale second.

15 The dating game

A hundred yards from Mam's back door was Kells Lane junior school, and we enrolled Sharron and Tracey-Emma there. Gavin went to my old preschool group, at the church hall on Cromer Avenue. Fortunately for us, the family home, New House, had four bedrooms. All my siblings had long since flown the nest, which left space for us, the invading horde. Although my mother initially asked me to help with the shopping and the running around that comes with having children, she quickly took on the lion's share of these duties. She was an absolute brick and I loved her all the more for that.

I thought then, and still do, that it was very important for the children to maintain contact with their mother. Carol was now living just outside Wakefield, about ten miles from Upper House. She very rarely saw the children, about once a month at best, and I did most of the delivering and collecting. At this stage there was no formal custody ruling, and I suppose I always assumed the children would be with me, since Carol appeared to have started a new life without them.

To drag the children to Upper House every weekend wasn't fair. It was several hours in the car and we would be arriving at a cold house. I had employed somebody to milk the cows during the week, but at the weekends they were my responsibility, along with all the other farm duties. I would have very little time to play and have fun with the children, so my mother child-minded for me while I drove like a madman down to the farm at weekends, to spend every daylight hour on a Saturday and Sunday on both the routine jobs and preparing the place for sale.

These weekends were quite lonely and gave me too much time to think, but occasionally friends would come by to see me and find out how I was managing. The most amazing visit came from a very attractive woman I knew. I had finished the evening milking and was making my way into the house when she arrived. I was in my green boiler suit with its all-important John Deere logo over my left breast, green "Compo" wellingtons and, courtesy of Hilda, an old bissom of a cow, a generous layer of now nearly dry cow poo halfway down my back.

I saw her as I opened the door to go into the house. She was dressed in a tight, red shantung silk shift dress and matching shoes with six-inch stiletto heels, and she was carrying an ice bucket with a bottle of champagne pressed into the ice and two glasses filled with bright red ripe strawberries. Just the right outfit for visiting a farm, I thought! I probably began drooling at this vision of loveliness, her shapely body swaying down the path towards me. With her full bright red lips she kissed me. I can't say she did it passionately, but it was a kiss that was full of the promise of more. She went in through the door before me and told me to strip. This was unreal, the stuff of fantasies. I couldn't get out of my boiler suit quick enough. All thoughts of manoeuvring my knackered arm into the correct position vanished. I had started with the wrong one and got myself stuck fast, half in, half out. I had also forgotten to take off my wellingtons first, but she stepped forward and said, 'Philip, it's all right. I'm here for the night, so calm down and take your time. Let me help you.'

Then, as if I was a precious thing she eased me out of my clothing. There I was in my front hall in my underpants and – *quelle horreur!* woolly socks pulled up to my knees. I stank of cows, yet she took me by the hand upstairs to the bathroom. I was unsure how to react, but again she took the lead and started running a bath. 'I'll be back in a few minutes to get you,' she purred.

She left, quietly closing the bathroom door behind her. I finished getting undressed, stumbling as I pulled off my sweaty socks and climbed into a sweet-smelling bubble bath. She must have brought the stuff with her, because I certainly didn't have any. I had been there for a short time, sufficient to wash all the places I could reach when, unbidden, she came in and knelt beside me, and instinctively washed where I could no longer reach. She also spent a long time massaging my stricken elbow, and, as far as I can remember, until very recently, that moment was the tenderest thing anyone had ever done for me. All I can say is she must have been watching from afar to realise what this meant to me. Then she pulled me out of the bath and towel-dried me, tenderly patting around my scars.

We made our way to my bedroom and I sat on the bed. 'My God!' I thought. 'When was the last time I'd changed the sheets?' I needn't have worried. She had changed them while I was bathing. She slipped out of her scarlet sheath, swaying to allow it to fall to the ground, revealing her exquisite shapely body. She was dressed in matching Janet Reger underwear, stockings, suspenders, red stilettos, and a smile that would have melted the Arctic. I stood up and took her in my arms, and I would like to be able to say that we went on to have eye-wateringly good sex, but I can't. Carol had had my virginity, as I hers, and I was so inhibited I was useless. But that did not matter because we laughed and talked, failed again, then laughed and just had a lovely uncomplicated night of enjoying each other's company and bodies.

There were no promises of love or commitment or anything in return. She felt that I was lonely, that life had been shitty to me, and I needed to be shown how to enjoy myself again. We occasionally saw each other out and about socially afterwards, but neither of us ever mentioned it again. It must be said that, whenever I saw her, that vision of loveliness flashed across my mind's eye and brought a long-lasting warm glow. Since then I have often thought about that cerebral dark pit I was trying to climb out of, and it wasn't a wonder pill that pulled me out, it was another human being doing the right thing. I am sure her then husband wouldn't agree, but, from my point of view, no amount of counselling, drugs or anything else could have done more.

Back in Consett, we were sourcing machinery for our premises on the Watling Street Industrial Estate, getting bottles and cartons designed, and setting up the laboratory. We had favourable inspections from public health, the fire brigade, and the health and safety people. Our logo was a cartoon cow that had been squashed into a box. Imagine a scene in one of MGM's *Tom and Jerry* cartoons, where Jerry shuts Tom in the fridge and, when he opens the door again, a square Tom slides out. Our logo was like that, but with a cow instead of a cat. Even finding customers was relatively straightforward, because we were willing to supply businesses that in the 1980s the big dairies wouldn't supply, such as garages, corner shops, mobile shops and small supermarkets.

The day arrived when we started employing people. Part of the deal with the council was that we took on young people or the long-termd unemployed. Even when the building was being erected we were swamped with enquires from people who saw it. Merry Dale acted like a beacon for the local population and, if my Saab was parked out the front, there would be a constant stream of job seekers, all asking for me. They were quite a mixed bunch, too. Our lab technician was a young chap with a string of 'O' levels and the title of Junior Mr Great Britain. He had muscles on top of muscles and his idea of a good gym workout was when he vomited two or three times. He had a permanent fake tan. These days it would be sprayed on, but back then he used some sort of stain the colour of oak. It looked seriously weird, but together with copious quantities of baby oil it helped win him his title. Having won the junior title he wasn't expecting to win another before he was fifty, until age had taken away his skin's elasticity and allowed striations in the muscle fibres to show through.

In contrast we had Punky, who could have been the model for Vyvyan Basterd from BBC TV's *The Young Ones.* He had some very strange ways and he mixed with even stranger people. One Saturday night The Rolling Stones were playing at Saint James' Park in Newcastle, a good 15 miles away. Punky and his mates were out on the moors above Consett, brewing magic mushroom tea. As the night fell and the strange lights on the far horizon from the Stones concert brightened with the encroaching darkness, the hallucinogenic effects of the mushroom took hold and they convinced themselves that we had visitors from outer space. *En masse* they tore down to Consett police station, and with the adrenaline from running heightening their terror they ran, raving about little green men, straight into the arms of the police.

The following morning Punky arrived at work, escorted by a police sergeant who, fighting back the overwhelming desire to laugh, explained that they had spent the night in the cells and had their collective backsides well and truly booted. He said that Punky was fit to work and he recommended a day's hard labour to sort out his upset constitution. I took him off the sergeant's hands, even though it was clear to me that Punky was still away with the fairies and the little green men.

Although my staff were all very different, team building was not a problem. I am very much a "hands on" manager and they responded well to not having specific roles. They realised that if the business succeeded, they would continue to have jobs, and if that meant putting your hand down a drain or picking up a spanner, then that was what they did. Dairy production started. Thanks to the way the team performed, we coped well with the teething problems encountered by any start-up business, and we quickly got under way. It seemed that barely had we begun, than the chilled wagon we had bought for deliveries was full every time it went out.

Around this time a loathing of modern communication firmly embedded itself into my psyche. It lasts to this day. The modern way of staying in touch in 1982 was the personal pager system, which involved clipping a device the size of what felt like of small plastic shoebox to your belt. When people wanted to get in touch with you, some sort of radio signal sent their telephone number to the pager, which displayed it on a tiny screen. I wore one of these stupid devices for three years and in all that time I never received anything that was so important that it could not have waited until I was back at home or at the dairy. The moment when it went into the bin was the day Mam, Granny Dixon's number came up on the screen. I was on the M1 returning from visiting either a business contact or an old friend – I'm not sure which, but I do remember I was in a hurry and the insistent tone of the pager drove me off the motorway and in search of a public telephone I could use. After visiting every vandalised booth in a 10-mile radius, which added three-quarters of an hour to my precious allocated time, I found an old fashioned post office with an internal public phone and managed to contact my mother.

'Nothing important son,' she said, when I'd got through to her. 'I just wondered what you'd like for your tea.' I knew from that point onwards that being in constant communication with the rest of the world at all times was a very, very bad idea.

At this point in Merry Dale's development I spent all weekdays away from the farm and with the children at New House. On the odd occasions when, at Mam's suggestion, I took an evening off, I would visit various Round Tables in the area. Being a member in one club gives you visiting access to meetings of all other Round Tables. At one of these, a smart ex-naval officer was giving an after dinner talk about his exploits in the brothels around the world. It was such a tasteless repartée that quite a number of the members objected to it. I was even more surprised, not to mention a little horrified, because I knew the man. It was the man I nicknamed Dick Dastardly, the business consultant that Mr Brown at Consett had arranged for me, and seeing him in this different role showed me what a shallow, twisted mind he had. I was horrified to think this charlatan of a man had in-depth knowledge of my plans.

Shortly after this incident two businessmen walked into my small office at the dairy. They introduced themselves as Mr Bob Elliot and Mr David Dod. They were wholesalers who owned a large number of milk rounds in North East Yorkshire and they announced they were in the market to buy 1500 gallons of bottled milk a day. Were we interested in buying this trade? They said it was worth around £75,000. We told them we were not interested in buying trade and thought that would be the last we would hear from them. A few days later they invited Noel and me to their depot and proposed a no-money deal, in the form of a partnership. It had not occurred to me that wholesalers as well as farmers would be interested in the project, so Noel and I returned to Merry Dale, and over the next few days researched their reputation (as much as was possible back then, before the advent of the internet), and we couldn't find anything detrimental. Noel and I naively agreed to a 50/50 share and invited them to join our board of directors. In hindsight, greed probably blinded our better judgement, but the deal represented double the capital input we had invested, and with that throughput of trade it guaranteed future profitability. We were blind to one of the great truisms; when things look too good to be true, they usually are.

Shortly after us setting up this new arrangement, the dirty tricks started. I have no clear proof of why they did, but I can only guess that it was because Merry Dale's arrival had started to become worrying news to the northern sector of the dairy business. The problems began when two people decided to break in one night and steal all the rolls of aluminium foil we used for our milk bottle tops. As luck would have it, our deliveryman decided on a 4am start that day, which was early for him, but because he was there at that time he caught the intruders in the act. I arrived on site to see these two, who were actually members of my staff, carted off in a police car. As they were being bundled inside it, one of them shouted to me, 'Don't worry Philip, we've left enough for a couple of days' production!'

And they had, because it meant that if they hadn't been caught, their jobs would have been secure. That's when I started to smell a rat. They can't have done it for the value of the goods. There was far more marketable stuff in the building, and who on earth would have bought their booty at a sensible price? It seemed such idiotic, mixed-up loyalty and it could only have been done for sabotage, to stop production for a day, so as to give us a name for unreliability.

As it was, the thieves never admitted to being put up to it by anyone, but my suspicions were heightened about outside interference when some of my suppliers suddenly refused to trade with me, not even for cash on delivery. Luck was on my side, because all our machinery was second-hand and a company dealing in scrap dairy machinery came to my aid. As for chemicals and dairy sundries, my old agricultural suppliers circumnavigated the embargo the suppliers were running.

As a footnote, I saw that one of Merry Dale's three-coloured bottles was auctioned on eBay at the time I was writing this book. It made £2.50. In 1982 it cost me 10p. Keeping a few thousand of them would have made a better investment than an Ashley Jackson painting.

But whatever troubles were looming on the horizon for Merry Dale, there remained the matter of disposing of Upper House Farm. In England there is a traditional process for selling a farm and giving up a tenancy. It's not the law; it's just the way things are done. First of all you set about finding a buyer or tenant and then you arrange for the farm sale, where everything that can be moved – livestock, hen crees, tractors, ploughs, buckets, cow chains, dairy equipment, old barbed wire, paint tins filled with rusty staples, etc is sold by public auction at the farm. When we bought Upper House, the Bamford brothers had long since given up full time farming, so they had very little at their farm sale that was of any interest to me. In fact I hadn't even gone to it. My sale was very different; I had a fully operational dairy farm with all the goods and chattels associated with it. A week or two before such a sale, everything has to be carefully laid out in long lines in the fields beside the steading, leaving enough space between the rows for the expected masses to examine the lots. The auctioneer arrives and values it all and lists it for the catalogue. Then, adverts are placed in the local and national press, giving a brief description and details of where to send for the sale catalogue. To my surprise, the catalogue was a substantial booklet, with over 400 lots. I was selling everything, from lot no 1, a heifer from my best pedigree cow that was in calf by AI to a prestigious bull, to lot no 421, a small roll of used barbed wire.

On the day of the sale the auctioneer arrived with his entourage and his mobile sale ring, from which he would sell the livestock. His bookkeeper set up in my kitchen, placing a table across the back door to use as a counter, from where she would receive cheques and cash from the successful bidders. From then on, everything was taken out of my control and I stood in a daze, listening to my distressed cows blaring at me accusatorially. I had not milked them the day before and their big, bountiful, overfull udders were getting painful. Cloth-capped strangers in scrubbed wellingtons demanded answers to questions I didn't listen to, and the auctioneer, his voice distorted by a mobile Tannoy system that was probably built by Marconi himself, gabbled his way through sixty lots an hour. Going under his hammer were my hopes, my dreams and what felt like my soiled underwear.

I had been under the fond illusion that the auctioneer's valuations were accurate, but what amazed me was that my highly prized, top line cows didn't reach their expected prices, but the buckets of rusty nails and rolls of rusty barbed wire went for silly money. That night, as I sank to my knees, surrounded by silence on the cow-soiled slope to the byre, I imagined my girls walking into the shippon to be milked. They would be so pleased to be doing so; they normally had no need of chains around their necks to hold them in place. I openly wept for the loss of my girls and my dreams. Anyone who has lost a close pet can start to appreciate the grief, but mine was sixty-fold. Many of my girls I had eased into the world, the baler band ever present in my pocket to help a difficult calving. I had nursed, fed, cleaned and trained them, and in return for this partnership they had given me their milk, the income from which I provided my family with a home. Now it was no longer mine.

Filthy dirty, I climbed into my Saab and drove slowly back up north to my old family home. Because I was feeling emotionally very unstable I suppose I expected a bit of sympathy from Mam. Instead, she took one look at me and said, 'don't you dare! What time do you call this? The sale was over hours ago, the children need their bedtime stories and what shall we have for tea tomorrow? Can you take Gavin to play school tomorrow before you go to the dairy? What day can we go out clothes shopping with all three? The rags you call clothes may have been suitable down on the farm but they're not suitable here...'

I don't think she let up for fifteen minutes solid. It was just what I needed, because I had to face reality. I had three gorgeous children to raise and we needed to get a life going again. My mystery lady visitor had got one thing right – I was lonely, and I needed some female company. But how do you ask women out for a date when you're nearly thirty years old and had never actually been in the dating game? It seemed incredibly difficult, if not impossible. Carol and I had been together since we were fifteen. Our divorce was something of a non-event; just a signature on a piece of paper and it was all over. I don't recall any wrangling about money, probably because there wasn't any and certainly no discussion about the children. I had always taken it for granted they would live with me. I certainly didn't feel like much of a catch with my knackered arm, three children and Merry Dale now my only source of income. All the good looking women seemed to be spoken for, or had so much emotional baggage they would have created a strike with the handlers at most airports. And anyway, I wasn't in a fit state to cope with other people's problems.

The other thing that I was about to discover and was a complete eye-opener for me was that women of my age had as big, if not bigger, sex drive than me. This wasn't because I was some macho man who thought women should stay at home cooking my tea. No, it was because somebody, I have no idea who, had indoctrinated me to believe that men were just lusty bastards, ruled by their one eyed trouser snake and women were gentle creatures who needed to be wooed, and then after a decent amount of time might allow you, as a self-sacrificing, loving gift, to seduce them. How is it that as young people we could have grown up with such enormous gaps in our education? We all knew the capital of France or our six-times table, but the emotional stuff we have to learn ourselves. Then having made our mistakes, we get conflicting advice coming at us from out of the woodwork.

It was with all this mental debris I entered the dating arena. I think my first date was with a twenty-eight-year old blonde petrol pump attendant who worked at a filling station I used on my weekly commute. I had gone for her looks, and hadn't considered why she wasn't already in a relationship. We had chatted several times when I had bought petrol and I commented on a poster for Ballet Rambert in her little cubicle. She said she loved them and wanted to see them, but didn't have anybody to take her. The performance was the following Saturday at the Theatre Royal, Newcastle. Not grasping how I had been manipulated, and being forever gallant I volunteered for this far from onerous task. I was quite chuffed at getting myself a date. It wasn't only my first for fifteen years, but my first ever date as an adult. And not only that, she was very tasty too.

We had exhausted our conversation within about five minutes. Normally I can talk the hind legs off a donkey; actually, make that a herd of wildebeest, but you need some sort of feedback even if it is just, 'yes' or, 'I see,' or something. But there was nothing. It was like trying to extract conversation from a wet lettuce. As for the ballet, I have always hated live ballet, not that the two performances I've seen make me a world expert on the subject. I much prefer it on the telly, where you can't hear thirty graceful swans crashing onto a hollow wooden stage, sounding like they've got clogs strapped to their toes. The thundering sound completely blows away the illusion for me and, unfortunately in this respect, Ballet Rambert was no different. And it was a bit bizarre compared to the classical stuff. Needless to say I did not see her again and changed the garage I used for my petrol.

I had kept in touch with Ron and Claire Sorkin, the London couple who had rented one of the cottages at the Coles' farm. Ron had left Redicut Carpets and gone to work for another mail order company. They had moved back to the big smoke, or rather, Bushey, in Hertfordshire. All those open spaces and lack of people in West Yorkshire had given them chronic agoraphobia. They'd sent me several, sometimes very insistent, invitations to visit them, and in those days, when the police traffic patrols were few and far between, a journey there took just over three hours.

Free time is a novelty when you are a dairy farmer – it's a 365-day working year, and you can only ever switch off if you leave the farm, because there is always a mountain of work just outside your front door. Even if you have already worked sixteen hours that day there is an unstoppable urge to go and reduce that mountain. Proper dairy farmers never join the dating game and never have affairs. It's simply a matter of logistics: you do not have the time. After you've had your supper and sat down on a comfortable chair to have a chat with your loved ones, your head is lolling within five minutes, your mouth wide open, you are snoring like a chainsaw and probably dribbling out of one side of your mouth. It's no wonder that most dairymen are divorced or unmarried, and the only long-lasting relationships are wives or partners who haven't passed their driving test, or are as foolishly obsessed with cows as their husbands. So, having free time was a complete novelty to me.

While I was setting up Merry Dale, the dairy only operated five days a week, so I had time to drive down south and visit the Sorkins. While I was visiting them I got to know Marjorie, a close friend of Claire's who I had met briefly once before when they lived next door on the farm. Marjorie was a very slim, elegant, blonde lady, and a whirlwind of a woman. Why she didn't have a queue of suitors was beyond my ken. She was a beautiful person to whom I owe a great debt, we were an item for many months: she would have made a lovely step-mum for my babes. She fell for me hook line and sinker and for the children, too, and we spent a lot of time together. With the harsh reality of hindsight, she was the person I should have entrusted my young children's safety and lives to. Her love for me was incredibly seductive and I think that was why it took me so long to realise that, although I liked her a lot, I did not love her, and life had taught me one-sided relationships ultimately do not work. When I called a halt to our relationship, much to my shame, I hurt her so deeply that she changed her life completely and joined the Royal Army Nursing Corps.

16

A second marriage, and life in a castle

Like my brother Noel, my sister Jacqueline is a pharmacist. My other sister, Katharine is eight years younger, and, around the time Noel and I had started Merry Dale, she was completing a biochemistry PhD at Leeds University. Katharine, ("Stinker" as you will recall I affectionately used to call her), lived the wild, exotic life of an eighties student, with all the excesses that university life brings with it. She lived in Woodhouse, in a Victorian terraced house with, officially, five other people but with partners coming and going, scheduled and unscheduled sleep overs, it was a very transient household and you were never quite sure who was going to walk through the front door. All six tenants had their own rooms, with Yale locks on the doors to give them privacy. Some had big rooms and some had tiny attic cubbyholes, but they had a communal bathroom, kitchen and sitting room. I used to call in to see Katharine regularly in her first floor, poster embellished, girly space when I was *en route* between Upper House and Consett.

When I knocked at her front door it would invariably be answered by some well-spoken, unwashed, dishevelled young slegger. Katharine was not often there because she had an evening job as a barmaid at The Chemic, an extremely busy spit-and-sawdust real ale pub, just 500 yards away. The pub seemed to be used as the household's sitting room, which would be no surprise if you had ever tried sitting on the sharp ends of the exposed springs on the "comfortable" chairs in their front room, always supposing you could make it across the sticky brown carpet without losing your shoes. The communal areas in the house were grim. The kitchen was completely skanky, and with empty takeaway cartons festering in amongst the damp, fusty basement it looked and smelt like an abandoned Kolkata street market. Katharine blamed her male co-tenants for the state of the place because there was no way she would act as a skivvy for them.

One evening I was sitting in The Chemic, drinking a pint of Tetley Bitter that Katharine had pulled correctly, in the correct sized glass that allowed for the extra volume of tight creamy head and no frogs' eyes (bloody hell, woman, just let me drink it!). Apart from a few ancient diehards, The Chemic seemed to be the preferred haunt for all the serious academics and medics from the university. They gave the pub a really scientific atmosphere. Even all the bar staff were doing PhDs and one of them in particular caught my attention. You could hardly miss her, because her hair was curly, died red with henna and sculptured into a giant quiff. Her name was Jean and she wore a bright yellow tee shirt over tight, royal blue trousers and a pair of bright red ankle boots. Was she pretty? No, but she was startling, talkative and completely extrovert. The surprise came later that evening, when I walked Katharine back to her room, because Jean came with us, and although I had never seen her at the house, she was one of the co-tenants. She was also a close friend of Katharine's.

Over subsequent visits we became friends and started seeing one another. She introduced me to university life, with its frequent and varied faculty get-togethers. We did pub-crawls, discos, fancy dress parties and those dos where you stand around holding a plate and a glass, supposedly having stimulating conversation. It seemed to me there was an awful lot of playtime amongst the work. Jean was investigating the effect of drugs on the bugs that cause diarrhoea and she had a whole host of bugs and human kidney cells to keep alive. It was a bit like livestock farming but without the tractors. Her livestock lived in Petri dishes and rotating conical flasks and needed feeding every few hours to keep it alive. I found it all fascinating. It was at some point in this time of my mixed-up new life that our relationship turned from friendship into courtship. I am not entirely sure why I fell for her. Perhaps it was because her life was so different or on the surface seemed to be fun, but before I knew where I was, we were getting married and I needed to move out from Mam and Dad's and find a house for us all to live in.

After scouring the countryside to find something we liked, could afford, had space for my family and of course my beloved Suffolk baker's table, we came across an apartment for sale in Stanhope Castle, in the village of Stanhope, County Durham. Before a local builder converted it into various flats it had been an approved school. The flat we bought was only the third one to be sold. At some point in Queen Victoria's day the castle's internal courtyard had been turned into a large solid-roofed orangery, with four south-facing, fifteen-foot high arched French doors that opened out onto formal lawns. At the front of the apartment, framed by the French doors, there was an enormous spreading beech tree, so I christened the flat "The Beeches." This area was basically the footprint of our new home.

The French doors took up one wall of the dining and day room and kitchen, giving them a conservatory-cum-garden feel that is very much the mode today. At the back of this room was a large, elegant oak door that led to what had been the castle's oak panelled dining room and it contained a marble fireplace that looked big enough to burn whole trees. The oak floor had been polished by generations of the former residents and had a patina that would have sent TV antique specialist Tim Wonnacott into fairyland. There were three bedrooms and a galley kitchen, all new and ready for us. Joint ownership of five acres of woodlands and formal gardens came with the deal. There was no formal arrangement for maintenance; I didn't need one, because The Beeches was sold freehold and the way it was situated within the building, there was nobody above or below me. Access could not have been more convenient, as it was situated at the head of Stanhope High Street. The village, despite being in the middle of nowhere had all that we required for daily life. The village even had its own gas supply, tankered in on a weekly basis.

Jean and I got married in Stockton registry office on a gloriously sunny spring day, with just a few close friends and family. We had our reception at The Beeches, with the big French doors wide open to take advantage of the weather. The catering was on an organised faith supper style with all the guests bringing a contribution to the wedding breakfast. I supplied the drink, including a barrel of beer, and the bar was on a self-service basis. I have always loved family parties when all the children are in their finery, playing together with distant cousins and their friends. This reception took this to the extreme, as there were as many children as adults and everyone had a ball. Noel was entertainments officer and he organised treasure hunts in the grounds and party games adults could join in with. This was a very deliberate decision so that Sharron, Tracey-Emma and Gavin would feel the day was for them too. After all, Jean was effectively marrying four of us.

The village school was within walking distance, and like all village schools in the 1980s, the small number of teachers taught a wide age range, but it still managed to achieve a high standard. Gavin was even taught computer maths, which is quite remarkable when you consider that basic word processors were about all that most schools got.

With all the positives of living in the high end of a dale, there had to be a downside, and it came in the form of the two roads I needed to use to get to work. One followed the River Wear to Bishop Auckland, but it was 40 miles longer than the other, which went up and over what was then the highest metalled road in England. Snow lay on the high points from November through to April, which made for interesting winter driving conditions. Two years previously I had bought an automatic Saab. It had a hatchback and cavernous luggage space, which I could use for delivering milk in an emergency. I had chosen it because it was the safest car on the market at the time and was supposed to be brilliant in snowy conditions. With that in mind, I always drove to work using the high road.

While I was based in Yorkshire it drove very well, but by the time we moved to the North East it was becoming decidedly sluggish. We delivered milk to a small industrial unit on the other side of Consett, to a company called M.A.D. Developments. They ran six cars that took part in the RAC Rally. I was bemoaning my car problems to the director one day and he told me to bring it round to him that afternoon. He met me in his car park and he was carrying the smallest, poshest looking toolbox I had ever seen. He told me to leave the engine running and to slip the bonnet catch. When he opened his miniature toolbox, much to my surprise, he pulled out a stethoscope and proceeded to listen to various parts of the engine. He shook his head, as all mechanics do when delivering bad news, and my heart sank.

'You'll have to leave it with me,' he said.

'How much will it cost?' I asked.

'How about free milk for a month,' he said, 'if I make a difference?'

'Are you just going to tune it?' I said. I knew the value of the milk, which wasn't exactly colossal so I was trying to figure out just how much or how little work he planned to do.

'Oh, this and that,' he said. 'I've a few bits lying around that might be useful.'

I went to collect it the following day, delivering his two free pints of full fat. From the outside, my Saab looked the same, and when I started the engine it sounded similar, although perhaps a bit throatier. But when I pressed the accelerator, I unleashed a beast. The speedometer was calibrated up to 140mph; the car had reached as much as 105mph when it was at its previous best. At 2 o'clock one morning on a deserted Durham bypass I managed to badly bend the pointer at the 140mph stop point. I hadn't a clue what speed I got up to because I didn't dare take my eyes off the road and it was only when I returned to normal speeds did I discover the bent needle. I was OK at those speeds as long us I kept my wits about me. The police, at that time did not have the equipment to catch me.

I abused the Saab dreadfully. When we had developed a big enough round to fill our chilled wagon to capacity I developed a wholesale round solely for delivering Pergals of milk. These are 3 and 5-gallon plastic containers, a bit like big wine boxes, that fitted into the commercial milk dispensers that cafes use. The old girl finally went bang and pushed a piston out the side of the engine block one day when I had 700kg of Pergals in the back and I was doing 60mph up a 1 in 6. What a car!

Our life as working parents settled into a normal routine. We got to know most of our neighbours, who shared some of their adventures, both past and contemporary with us. We had secured the services of a local child minder, a young mum called Katy, who looked after the children before and after school. I had a very early start, so Jean normally delivered them to her before she left for her long drive into Newcastle. Jean was employed at the University, researching into something to do with retina attachment and when she got home she still had her thesis to write up. Following her three years post-graduate studies at Leeds University she had a mountain of results to sort through, analyse and come up with some new scientific conclusions. Then, if she presented them in an oral examination to a panel of professors and academics working in that field (it's called a "viva," from the Latin, viva voce, or "live voice") and if they agreed with her results she would be awarded a Doctorate (PhD). This process took her about two and a half years of all her available spare time, but after overcoming some analysing problems she secured the prestigious qualification and was allowed to become Dr Jean Bennet (all her research was done in her maiden name).

I tended to get home first, so I collected the children from Katy. This worked really well apart from the day when Katy was not able to collect the children from school and we had arranged for Rob, a friend of mine, to do the honours. I turned up at his house on Stanhope High Street at four o'clock and he invited me in for a natter and a cup of coffee. My children were playing in the front room with his two children. They went to the same school; his son was in Gavin's class and his daughter was in Tracey-Emma's. Rob and I were sitting at his scrubbed pine kitchen table, half way through our coffee, when he suggested that we should sample his new batch of home brew. So we did. I have a vague recollection of drinking half a pint of a really nice beer, but my next memory happened about four hours later, when I was lying flat out on the settee in our sitting room in The Beeches with Sharron kneeling beside me with a cold flannel, mopping my brow. She was speaking to Jean who had just got home and was standing in the open door. 'Daddy is very poorly,' Sharron said.

Much to my shame I was as drunk as a skunk. I had even driven the children home, because the car was parked outside. I had no recollection of this at all. The next day, as it happened to be a Saturday, I had a bit of a lie in and at about 10 o'clock Rob knocked on the door. He asked if I had a bad head, because his was stotting around like a puck at an ice hockey match. I was holding my head up with my working arm to preventing it from falling off and bouncing around on the floor. Without another word he came in and went up to Jean and said, 'the wife has sent me around to apologise. She says I was a bloody fool and nobody should suffer my evil potions.' With that he turned round and walked out. Unfortunately, his apology didn't prevent me from being in the doghouse for several days.

The gatehouse to the castle was a Hansel and Gretel-like structure with tiny, fixed leaded windows. It was occupied by a couple called Jane and Bill. One night at about 11.30, just as they were drifting off to their respective lands of dreams, undisturbed by the noises that an ancient building makes when the timbers contract and give up the heat of the day, they heard a new noise, a gentle fluttering of wings brushing against their bedroom window. It made the loving couple pull the duvet up a little higher, but halted their drift into the land of nod. Five minutes passed and oblivion once more settled on the pair. Then there came a funny little cooing noise from the outside windowsill, which brought them wide awake. They both turned their bedside lights on and sat up, looking at the tiny curtained window. It was clearly not a burglar or anything dangerous, so the brave duo got out of bed, made their way to the window and drew back the curtain.

Sitting there on the narrow sill, with his back to them, was a Little Owl. The miniscule chap was preening himself and making contented little cooing noises. Jane and Bill thought he was lovely; they smiled at one another, closed the curtain, switched off the light, snuggled into the duvet and cuddled up to one another. Sleep descended very quickly on them, but not for the ball of feathers sitting outside, for he had found his ideal hunting perch.

The lights for the castle grounds were in front of him, giving him a clear, shadowless killing field. For the next hour or so he filled his near-bottomless belly with all the shrews, voles and dormice that ventured out from the surrounding woods in search of food. This miniature raptor took each catch back to his window ledge to hungrily devour. When he had finally had his fill, he decided that his position of superiority in this area of Weardale needed establishing. He puffed out his chest and gave a long loud hoot, letting all his rivals know he was the chief here. He felt one hoot was not enough and he continued until Jane and Bill could take no more. Bill threw back the covers and stormed over to the non-opening, antique, listed, leaded window and rapped as hard as he dared. The owl stopped hooting and did that peculiar owl manoeuvre where he turned his head 180 degrees without any other part of his body moving. With his unblinking, enormous saucer shaped eyes, he looked directly at Bill, who was flapping his arms near to the window, shouting, jumping up and down and doing anything short of smashing through the window and knocking the fluff ball off. Basil Fawlty could not have given a better display of frustrated rage. The owl was totally unimpressed; he looked disdainfully at Bill and turned back to gaze at his hunting grounds. It was at this point Jane started to snigger. Bill realised how ridiculous he looked, prancing naked in front of the curtainless window and he started to laugh as well. Their tears of laughter turned to tears of frustration when, four hours later, the owl was still to-wit to-wooing on the sill.

The next day they were so annoyed they told everyone they met about their night visitor, but in doing so negated their chances of using more violent methods of ridding themselves of the legally protected night hooter. They had no idea it would go on intermittently for months. They tried every method they could think of to stop the pesky bird landing on their sill, before finally vacating the only double bedroom and moving into the box room down the hall. Their efforts included Bill standing in the courtyard at two o'clock in the morning in his dressing gown and slippers, the cold night air whipping round his nether regions, trying to poke the little bleeder off his perch with a fifteen-foot bamboo pole. As soon as Bill returned to his bed and turned off the light, the owl would ruffle his feathers, chunter disgruntledly under his breath and return to his perch.

It was at times like these that the use of a shotgun seemed the best and only solution, but that was out of the question, as the owl was protected by law. Of course at three o'clock in the morning, the temptation to break the law was attractive, but the damage it would have caused to their cottage would have been prohibitive. Another, legal method they tried was installing those spiky strips that they use to keep pigeons off ledges in city centres. The problem was that owls are far more intelligent than pigeons. Using nesting technology, their little avian professor filled the gaps between the spikes with sticks and moss to make a substantial platform from which to launch his nightly forays.

Then, one Friday night in winter, there was complete silence from the windowsill. Bill and Jane enjoyed a full night's sleep and a three-hour lie in on Saturday morning, followed by a slow, leisurely breakfast. To blow the cobwebs away on this sunny, crisp clear morning, they decided to take a walk in the woods. Putting on woolly hats, scarves and gloves, they set off down the path. They hadn't gone fifty yards when they saw, in the middle of the path, frozen stiff, dead as a doornail, the Little Owl. Bill wanted to stamp on the little bugger with glee, but Jane went all gooey about it, picked up the little chap up and took it home. She had him stuffed, making him a permanent member of the family. It was the ideal compromise – Bill had the satisfaction of having the job he wanted done in the first place, and Jane had a little feathered memento for her mantelpiece.

Derek and Joan occupied a flat around the corner from The Beeches, above the castle's old wine cellars. They had moved there after living in the Raffles Hotel in Singapore for the previous twenty years. Derek was the wine correspondent for Diners Club International's monthly magazine and had a wine collection second to none. It was all stored in neat, catalogued rows in the stone-vaulted cellars, which protected it from the harsh weather conditions of upper Weardale. Derek had been a naval officer before taking up his colonial lifestyle with his young bride in Singapore, and I'm afraid that they did not live in the real world. They'd come back to England to retire, and on disembarking in Portsmouth to start their retirement, were dismayed they couldn't fulfil their dream return to Blighty: go into the nearest pub, order a pint of Guinness and a dozen oysters each. They could have got the Guinness, but not the oysters. It was fish and chips or nothing!

One cold December morning Joan arrived at our door to introduce herself. She was dressed in a baby pink cat suit, fully made up and sporting a blonde bubble perm. She looked like a Eurovision song contest entrant from the early seventies. 'Where,' she asked, 'can I order a daily delivery of fresh flowers?'

Tact is not one of my strongest attributes, but in an effort to be welcoming I suggested some creative foraging in the castle's gardens and woods. She was more likely to find Lord Lucan in the local shops than fresh flowers. She was obviously a lady who was well used to getting her own way and she pursed her lips, turned on her heels and said, 'Well, I'll have to see about that. Thank you!'

How she managed to make her thank you sound like 'You stupid man,' I'll never know, but it must be a generation thing.

This pink apparition became a regular feature around the castle as she busied herself trying to make the grounds look presentable, but unfortunately she was totally ineffectual, because she flitted from task to task, never quite getting to grips with anything. One snowy morning just after Christmas, I became aware of how quiet it was in the house, and realised the children must be outside. Pulling on my wellingtons, I went in search of the mischievous trio and when I rounded the corner of the castle I saw, on the doorstep of Derek and Joan's house, three small pairs of boots, all standing in a neat row. I had never been inside their house, but imagine my dismay when I was shown into the sitting room. It was furnished with a light cream carpet and a large, matching three-piece suite. There, sat in the middle of the practically white, sumptuous settee were my fun-filled three, grinning from ear to ear, each cradling huge mugs of steaming hot chocolate.

It was one of those moments when the art of Zen becomes paramount. I sat in an armchair opposite them, and with well-chosen, calm-filled words prayed silently, because no matter how confident you are about your children's behaviour, hot chocolate and white furnishings do not mix. Joan flitted around the room like a fussy, emaciated pink chicken, titivating the ornaments here and plumping cushions there. After an incredibly tense quarter of an hour for me, but a good time for everyone else, I was able to extract the little darlings without any of the children depositing chocolate stains on Joan's white furnishings. As we were leaving, I hung back and I explained to Joan the risks she had taken with her pristine interior. I was amazed how she waved my cautionary words aside, saying they were welcome to a drink of chocolate any time they wanted, and on her cream sofa. That said, I can't remember her ever having them visit again.

In 1986, the snow fell heavily and the valley sides and minor roads became brilliant sledging tracks. Sharron was eleven, Tracey-Emma was seven and a half and Gavin six and for Christmas I bought them each sledges from Fenwick's in Newcastle. I had never seen the like of it, neither before nor since. They were like a narrow, solid Li-lo, and if you put "snow body boarding, air boarding" into YouTube, you will see something approaching their mode of operation. They could take three people sitting or one person lying down. They would slide on wet grass, let alone a snowy slope, and were faster than most skis. Obviously I had to choose the slopes carefully so that the children didn't get into trouble.

One Sunday, my parents were visiting and Dad, Grandpa Dixon, was keen to have a go on one of these sledges. Grandpa, you'll remember me saying, was not a practical man. Granny (me Mam) would dread the days he would attempt any DIY. In fact she would hide his tools so he couldn't use them. The damage he caused was always costly and the day would end in arguments because of his frustration at his ineptitude. His driving skills were no better. He always drove big, 4-litre models and they would become bashed and dented very soon after they left the showroom. He would approach a roundabout thinking he would be turning right, and sure enough he would, but he'd drive the wrong way around it. In his mind, that was a more direct route. One day he drove round a corner, straight into the bucket of a big bulldozer. The list of the bashes and scrapes he got into is endless.

As we stood on the children's sledging track together, Grandpa kept on looking over at where the village's teenage bucks were having a raucous time, sledging, skiing and sliding down a precipitous run. He turned to me and said, 'That track looks good fun.'

'No, Dad,' I said. 'These sledges go too fast. Go on a gentler slope until you have learned how to control it.' Of course, to Grandpa, this was a challenge to his age and his masculinity, so off he went, trudging his way through the deep snow to the top of the hill, where the boys had compacted the track into an ice run. Halfway up he had got hot, so he had unzipped his waterproofs and put his bobble hat in his pocket. He paused at the top only to wipe his fashionable (at the time) over-large spectacles, and then threw himself forcefully onto the sledge. It was like watching Eddie the Eagle – even the glasses were the same – and we all expected disaster, even though we could not help but admire his courage.

Man and sledge careered down the hill at breakneck speed. At one point, the two of them separated in mid air over a large bump, but managed to reunite just as the sledge crashed into the ground and continued their journey towards terminal velocity. The whole journey couldn't have lasted more than sixty seconds and Granny didn't know whether to laugh or cry. Grandpa overshot the end of the track and was heading at the speed of sound in the direction of the River Wear, but the impending cold, wet dip was fortunately avoided when he flew head first into a large snowdrift. All we could see were two snow-covered legs sticking out. We all rushed over and pulled him out. This was hard work, because even his wellingtons were full to the top with impacted snow. Rolling him over was just as hard, because his partly unzipped jacket had filled with hard snow. His head was caked in snow, but somehow his glasses were still visible on the surface of his snow-white head. As he shook off his snowy casing he laughed, and laughed, and laughed, making him look like a vibrating comedy snowman. The happiness virus had infected us all and tears of merriment coursed down our cheeks.

When he had gathered his composure he started out with determined strides for another go. Granny ran after him and shouted at him that he was to do no such thing. He continued on his way, leaving Granny behind. Bad move. You don't ignore Granny. She chased after him and with a flying rugby, knocked him of his feet. This started us all on another round of uncontrollable laughter.

The Cows' HIll Hotel

Walton Siddle should have been an actor. Or a comedian. Instead, he was the landlord of The Cows' Hill Hotel, which was perched in the High Pennines, in the upper reaches of Weardale on the cusp of Durham and Cumbria. It's one of those wild places where curlews cry down to the deep covering of heather and the sheep shelter behind crumbling grey drystone walls that were splashed with purple fluorspar and yellow lead sulphate.

To say the bar at Cows' Hill was old fashioned makes it sound too modern. It was ancient. It's the sort of place where Wackford Squeers would have gone to escape the daily routine of running Dotheboys Hall. The floor was stone flagged, the low overhead beams had globules of congealed nicotine nestled into the grain of the dark oak and the furniture would have had antiques expert Arthur Negus waxing lyrical for hours. The actual bar was no more than an enlarged servery, a metre-wide hole cut in the wall with two wooden doors that stayed open all the time. It was the sort of facility you'd find between the dining room and kitchen of a 1930s semi. The room behind the doors had vaulted brickwork of the kind you'd expect to find below ground level and there, nestling on their wooden trestles, were oak firkins of beer, one or two already tapped, the rest waiting to settle after their long trip up the dale. Leaning on the bar in front of them would be Walton.

He was stick insect-thin and had that unfortunate skin pallor that made you think he could do with a good bath. His face was covered in red thread veins, a consequence of the fierce and constant northerly weather and too much of the brown nectar that rested in the oak barrels behind him. He was a *doppelgänger* for Albert Steptoe, even down to the bright kerchief (or a flamboyant cravat for more formal occasions) around his neck. Walton had a wide range of accents, from county posh to shipyard Geordie, which he used to maximise whichever character he was trying to portray. If a stranger came into the pub and approached him, he would give his "I'm as pissed as a fart" look. This, to the uninitiated, was totally convincing and it involved him peering out from under hooded eyelids as he wobbled and nodded his head. He would slur a cursory greeting to the stranger, and his regulars would know it was "game on." In serving the newcomer, he would be outrageously bad at his job, not filling the glass to its legal measure or pretending someone had given him £5 when in fact he had been given £10. He would have a sip out of your pint before he gave it to you and if you didn't complain he would go to the next phase, where he pretended to be even drunker. He would stagger around, his speech became incoherent and he would drop your change on the floor or into your beer.

His behaviour was so infamous that he was cited in the Campaign for Real Ale guide-book as a "characterful landlord." The real ale assessor must have visited Cows' Hill midway between beer deliveries, because when Walton's beer was good, it was fabulous, but for a few days before the new delivery settled it would be cloudy, and when it had aged a bit it would have been better on your fish and chips.

Once, I saw two young hikers, a couple, all Berghaus jackets and knitted bobble hats, coming into the bar quite timidly and sitting by the window. I was with a small group of Round Tablers, drinking bottled beer because the draught had not settled and was decidedly iffy. The male of the species, complete with thin goatee beard, approached in a manner reminiscent of Mr Bean. Walton was pretending to be asleep.

'Landlord,' the young man said.

No response from Walton.

'Landlord?' He ventured again 'excuse me.'

Still no response.

'Landlord!'

Again, no response. The visitor raised his voice a little. 'Landlord!'

Walton opened his eyes. 'WHAT DO YOU WANT? DO YOU THINK I'M DEAF?'

The visitor almost fell over backwards. 'No,' he said. 'I'm sorry, but could we have two pints of your finest ale please?'

Walton's eyes narrowed. Did he show a flicker of a smile in amongst the wrinkles? He turned his back and drew two pints via the old brass tap in the bottom of the barrel. The beer was decidedly murky, but the big girl's blouse said nothing as he handed over a £10 pound note. We all knew what was coming next; our illustrious landlord treated it as a £5 and gave him change for that. This cardboard cut-out of a man looked at the palm of his hand, where Walton had put his change. He gave a little whimper, picked up his pints and headed back to his young lady companion. She took a look at the beer, then at the change. In a loud whisper she pitched in to him like a Jack Russell worrying a rat and sent him back to put it right.

'Excuse me landlord,' he said. 'I'm afraid the beer is cloudy.'

'What do you expect?' Said Walton. 'Thunder and bloody lightning?'

Judging by the customer's stunned silence, this was not the answer he was expecting. By this time our group were audibly sniggering and he looked round at us for support. Instead, his other half was out of her chair with more than thunder written across her face. Walton had gone into his fake torpor by this time and was leaning up against the hatchway.

'Landlord!' She demanded.

Walton opened his eyes. She was not the gimp that her man was. In fact, with her cheeks flushed with fury, she looked quite hot! 'This beer is off!' She began. 'And what's ...'

A light switch went on inside Walton's body. All of a sudden there appeared a charming, debonair, (if a little smarmy) friendly landlord. 'I think,' he said, in his normal posh voice. 'I mistook that £5 for £10 and for that dreadful error, I believe I must offer you a complimentary drink of your choice. What would you like?'

As Walton presented her with her choice of drink, he leaned over the counter and whispered, 'the cloudy stuff will be good enough for Frank Spencer over there!'

She just giggled and returned to her seat with her drink, one of the cloudy pints and no extra change.

Weardale Round Table, which I had joined when we moved to Stanhope, met fortnightly at The Cows' Hill Hotel. Thirty or more of us squeezed into an upstairs dining room, where we were served superb "fruit of the local countryside" fayre, which various professional and not so professional local gamekeepers delivered to the kitchen door for a cash consideration. That upstairs room was mission control for all sorts of antics, and at the end of all our meetings we had a raffle. The prizes were normally trivial, such as a dirty magazine or a pint of Walton's not so finest ale.

One evening we had the area chairman visiting and we rigged the raffle so only he could win. At the end of the evening he was presented with a large cardboard box and told not to open it until he got home. Amazingly, he did as he was told and at one o'clock in the morning, in his pristine living room in one of the poshest suburbs of Newcastle, he opened the box. Out flew the biggest cockerel in the world, or so the chairman thought. Maybe it wasn't the biggest, but that cockerel was a big boy. It was dark in the box, which kept the cockerel quiet, but as soon as he saw the house lights, he thought it was time to get up and get to work. He flew to the top of their glass fronted display cabinet, shook his ruffled feathers, puffed out his copious chest and crowed at the top of his voice. Three things then happened. Their six-month old baby woke up and started howling, his wife shouted down the stairs what on earth was going on, and the chairman, balancing on the back of one of the comfy chairs in an attempt to catch hold of the avian intruder, tipped backwards and crashed into the china display cabinet. The next morning he was not Mr Popular with his wife. Then, to cap it all on the news the Ministry of Agriculture announced the non-movement of anything avian because of an outbreak of fowl pest.

Now any countryman would have just wrung the blighter's neck and set a coq au vin on to boil. He would have been tough, but tasty. However, this was the city and although their world was severely disrupted, the thought of wringing the bird's scrawny neck was not an option. Their next-door neighbour's redundant rabbit hutch was dragged out of the undergrowth and set up at the end of their garden and the odious rooster was put in it. A few days later, someone overheard one side of a conversation between a friend of the chairman's wife and the chairman's wife herself. It went something like this:

'How's your Dave's cock?'

'...'

'Aren't they always active in the mornings?'

'...'

'I can see that. The noise! It must have woken the kids.'

'...'

'And the neighbours!!! Do they not complain?'

'.....................................'

'Struts around sticking his chest out.'

But suddenly the complaints from the area chairman about this stunt went quiet, and because they did, our club was on full alert for some sort of revenge trick. We would find out why all too soon. Yes, we were all very grown-up about this sort of thing.

Around midsummer's day, Weardale Round Table would hold an annual tug of war competition. It was a very popular event, with about twenty other clubs taking part. We had some giants in our club and since it had begun we had always won, so rather than us winning it every year, the rules were sportingly changed so that the trophy was awarded to the team Weardale beat in the final pull. The venue for the tug was the ford that crossed the Wear at Stanhope. The ford was about 60 yards wide and across it were concrete stepping-stones. The rope was about 2 inches thick and 150 yards long, including 45 yards each side for the teams of ten to pull on. The teams would stand on the tarmac roads sloping down to the concreted approach to the ford. Normally there would be about a foot's depth of water in the ford for one team or another to be pulled into. At the beginning of the competition the rope was relatively light, but as time progressed it got heavier as it became waterlogged and it would sag into the river. When it did, the teams had to be reasonably strong just to beat the current and the weight of the rope, never mind the opposing team. There was a white flag tied onto the rope at the midway point. With my dodgy arm I was not expected to be a team member, so, every year, I volunteered to stand in the river to centre the rope, then give the all-important command, 'PULL!'

On the tug-of-war event following the cockerel stunt we played on the visiting area chairman, we set up the beer tent, the barbecue and the car parking the night prior to the competition, all of which were on the Stanhope side of the ford. When we had finished we went off to the pub. What we didn't know at the time was that when the coast was clear, the area chairman's hit team arrived like the SAS on manoeuvres, tied one end of a rope to the tug-of-war rope and the other end to a bollard on the far side of the ford and vanished back into the night.

At six o'clock the following evening, when the first heats were taking place across the ford, all was progressing normally when, late as usual, the area chairman's team arrived. When they reported for registration, nobody recognised any of them. The chairman was taking no chances about winning, because he had got the eight biggest men he could lay his hands on and had made all of them honorary members for the evening. They were enormous, but not in a muscly way. They were the sort of men who live on a permanent diet of pies and mash. Although a tug-of-war is a combination of strength and technique rather than sheer bulk, the chairman's team won all the heats, so they were the team that Weardale had to pull against in the final. They won the toss for who pulls from which side and opted for the far side of the ford. I took up my post in the middle and the area team crossed on the stepping-stones. Then a group of their supporters came across the ford, squashed into an Audi quattro cabriolet. It was considered good sport for anyone who took part to go as fast as they could past me, so that I got soaked from head to foot. The broad tyres on the four-wheel-drive Audi splashed more than most.

When they had arrived at the other side I asked both teams to take up the rope, then take up the slack. As the rope rose up out of the water, I shouted instructions to one side or the other to shuffle backwards or forwards until the centre flag was in position. Of course, no one knew the Pie Eaters had tied the rope that had been hidden the previous night to the end of the tug rope. They were so confident in the strength of the bollard they just leant on the rope and let the Weardale team pull their hearts out. Weardale did this very successfully, for not only did they pull the Pie Eaters into the river but uprooted the bollard and sent it clattering across the road. Hanging their heads in shame, not for cheating, but for getting found out, these dejected trolls slumped their way across the stepping-stones, back to the beer tent for several pints and large helpings of burgers.

The supporters in the Audi were left behind to clean up the bollard and untie it from the tug rope. They had to pass me again on their return journey, and without a doubt had intended to soak me as they had done on the way out. They were blissfully unaware that I had been splashed by over-exuberant drivers enough times to know this was going to happen. I kept a large yellow dairy bucket hidden beneath the surface, expressly for the purpose of revenge, but this was the first time I had had the opportunity to throw its contents over an expensive open-topped, white-leather-trimmed, deeply-carpeted Audi. They were drenched.

Another event arranged by us in the upstairs room at Cows' Hill was the annual Christmas shopping trip for the disabled adults who lived in the upper reaches of the Wear Valley. Between 6pm and 10pm one Thursday just before Christmas, Newcastle shopping centre closed for usual trade and opened its doors to the housebound population of County Durham and Northumberland, so that they would have independent access to the shops. This took place a decade before the 1995 Disability Discrimination Act came into force. I have no idea whether we Geordies were unique in this, but I wouldn't be surprised if we were. The Geordies gave up their playtime and turned out in their thousands to help lift, push and clear. They made sure nothing was inaccessible, no matter what the circumstances.

Most people in our group were collected in cars and transported to Stanhope to be transferred to a regular bus. If special buses existed back then, I don't know, but we didn't have one. Seats were removed from the bus to make way for wheelchairs; those who were able to sit were strapped into seats, not with safety belts but broad webbing tied to seat legs. Everybody was highly excited to be going out independently of family or friends. It was making it possible for them to buy small gifts to say thank you to their carers. They didn't care how we achieved it as long as they got there and back, even if they were uncomfortable for the journey.

One year, a chap who was in the final stages of a long drawn-out terminal illness got in touch with us. He was completely bedbound and his doctor advised that he should not be moved. Nevertheless, he was determined to go and he did a deal with the doctor; he could go if he didn't get out of bed. The doctor thought he had defeated him but he had not reckoned with Weardale Round Table in full flow. His bed was on the first floor of his house. Fortunately for us it was a single bed and his drip was on a long tube. We took out the window completely, heaved the bed out by using the top of the van as a platform and then manhandled it down to ground level, put bed and patient in the back of a van, together with four Tablers and boarded up the window of the house.

When they arrived at Grey's Monument in the centre of Newcastle, the doors were opened and the bed was placed on two axles so he and the bed could be pushed around. We pushed him into Marks and Spencer's, where he bought chocolates, brightly coloured hats, jumpers, food treats and slippers for all his family. Surprisingly, he did it all in about three quarters of an hour, so we asked, 'Where next?'

He answered straight away. 'The Hotspur, for a pint at the bar, and if possible, can I see it pulled?'

The landlord and his clientele were brilliant. The double doors were opened; one side of the doors had not been opened since 1832 so it was a bit stiff and needed some persuasion with a hammer, which materialised out of thin air. Tables and chairs were moved, an area of the bar was cleared and everybody lifted the bed above bar height for him to see his pint being pulled.

He downed it in one and then said. 'That was fantastic, man. Doc told me it really wasn't a good idea to drink, but bollocks to that. I'll have another, please.'

The Geordie mafia

The spring of 1986 arrived with a sudden burst. The fireplace in The Beeches was always hungry for fuel, so this was the best time to cut some firewood. Soon, the angry buzz of my chain saw, slicing its way through the snow-damaged trees in the grounds, was cutting through the stillness of the countryside like a swarm of African killer bees. I stacked the logs in our garage to dry out. This was not very convenient, because it was located in the lower courtyard, and I had to carry the firewood up two flights of steep, slippery sandstone steps, but it was the only place I could store it, dry and out of sight.

One cold, damp Sunday morning, Jean and I decided to cheer up the sitting room by having a fire, so I went down the stairs to collect some wood. The large, double arched entrance gates set in the castellated wall were standing wide open, which was unusual, because apart from my garage, this area of the castle was undeveloped and unsold. It was mainly used as a clothes-drying area because it was out of sight of the rest of the castle's occupants. I rounded the corner of the building to be further surprised. Someone had nailed closed my garage door with large planks of wood. As luck would have it, my toolbox was in the car nearby and, as I went to get it, I puzzled over who would do such a thing. It must have happened the day before when we were out shopping.

Setting to with my wrecker bar and claw hammer, I soon prised off the planks, although unfortunately I took off great chunks of wood and paint with them, which made it look a right mess. Then, just as I was opening the garage doors, a royal blue Rolls-Royce Corniche swept into the courtyard, its fat tyres squealing on the grey cobblestones. The driver's door swung open and Mr Angry got out. He was a swarthy, square-shaped man with no neck, and a belly that had digested too much pasta. His red, alcohol blushed face went white with fury and he bellowed, 'what the f*** are you doing?'

Diplomacy being my middle name, I replied, 'did you nail my garage door shut, you fat ugly toad?'

He then walked over to the washing line and said, 'this your washing?' Without waiting for an answer he yanked the clothes off the line, sending the pegs flying like startled swallows, he threw the damp garments to the ground and proceeded to stamp on them. At this point, I surmised he was an escapee from the local loony bin and I decided that humouring him would be best, but the sight of a real live Mr Angry having a full blown two-year-old-type tantrum in front of me made me laugh out loud. This just fuelled his temper, so he started to try and tear down the line itself, but unfortunately for him he wasn't strong enough and all he got for his trouble was two rope burns across the palms of his hands.

At this point, the children came out to see what all the noise was about and I moved forwards so that I was in between him and three curious little ones. I have always wondered why, in serious confrontational situations, people invariably back down before me. One day I am sure somebody will give me a fourpenny one or a "Geordie kiss," but true to my past experiences he retreated to the confines of his ostentatious transport. He gunned the engine and shot backwards out of the courtyard and straight into Stanhope High Street. Fortunately, no one was in his path because there is no way he would have been able to stop in time if there had been.

As we were gathering up the soiled washing, the village post lady, always keen for a bit of news, stopped on her round and asked what had happened. When we told her, she said, 'that'll be that Marra feller. Nasty piece of work, that one. Rented that house along the High Street. Watch out for him!' And with that she went on her way.

It wasn't until the following Tuesday, in the upstairs room at Cows' Hill, that my Tabler friends enlightened me about him. His real name was Angelo Marazzo and until he absconded to Spain after his business partner had been found, riddled with bullet holes in the back of his Aston Martin, he had been the slot machine king of the North. It was said that he was more powerful than the Krays and it was Reggie and Ronnie that took over his North Eastern business when he had his fifteen-year holiday in Costa Criminal.

The next contact I had with Mr not-so-big-now-but-bloody-dangerous-nevertheless-Angry was when I was trying to get the old castle gang lawn mower going. I was bent to my labours and had just skinned my knuckles when a none too friendly voice said, 'Put that back where you got it from, NOW!'

I turned to see a familiar square frame. 'What's that got to do with you?' I asked.

'I own the bloody place,' he replied.

'That means we must be partners,' I said. 'Because I do too!' I could never keep my mouth under control.

At that, he turned on his heels and strode off across the lawn, back to his blue gas-guzzler. Later, and much to my dismay, I was told by the building contractor who had sold me The Beeches, that Angelo had bankrolled him to enable the development. The poor builder had been well and truly bollocked by Angelo for selling me my property freehold. All the others in the complex were leasehold and what's more, to add to my delight, he was going to be a close neighbour of mine, as he was moving in next door!

I was dreading the day he took up residence, but the day came and went without any incidence and he was perfectly civil to my family and me. When a man like that moves next door to you, well, you worry, even though you feel right is on your side. That concern was not diminished when one day, while we were having lunch at the baker's table, the blue Mafia-mobile slithered sideways to a halt outside our French doors, the driver's door swung wide open as if it had been caught by a hurricane gust and Angelo fell out onto the tarmac. Clearly, he was not well. I told the children to stay inside and I went out to see if I could help. He was clutching his side and appeared to be in considerable pain, and by the time I had got to him I saw that there was blood oozing between his fingers. He said he had been stabbed (as you do) and could I help him to his flat? Much to my surprise, even though he was a bulky chap, he was not strong and needed a lot of help to get home.

I knocked on the door and his wife, a good twenty years his junior with their baby on her hip, opened it wide and he staggered in. I asked if he wanted me to phone for an ambulance or get our local GP, who was a fellow Tabler, to help. Angelo was adamant that he didn't need help. Do nothing, he said. He would sort it out. That night, as I lay in bed waiting for the sandman to do his work, Mulling over the day's events, I thought, had I really helped a Mafia don, and in some small way earned a "debt of honour?" Was it so easy to be drawn into the Cosa Nostra? Surely not.

Consett was known locally as the Red Town, because the red spume that used to belch out of the steel works stained everything it touched. At night, when the furnaces opened to pour the pig iron into their casts, the red hot flames and sparks lit up the night sky. Anyone who saw this, even from as far away as Newcastle said it was like watching the gates of hell, with the skyline opening up and spewing the devil's waste onto the town.

Next door to Consett was Stanley, a coal-mining town that nestled amongst the "slag mountains." The coal dust from the mines that had hung over the town for a century had permeated the brick and sandstone buildings and turned them all a uniform black. This wasn't a Hovis advert version of the North; this was real. The men of the town, filthy with coal dust, returning home after a hard day of back-breaking grind, would cough up clumps of pneumoconiosis sputum – gold watches, they called them – onto the blackened pavements. Except they didn't go home straight away. In between the two towns were the public baths. These were not swimming baths, but facilities for the miners to get clean and ease their knotted bodies after they had expended more calories in their toil than it was possible to consume in a day.

This was a testosterone-driven place for hard men, a hangover from the days of the industrial revolution, but the first day I visited I was staggered. It was fabulous, like stepping into some sort of Moorish palace. You almost expected to see a shapely Turkish belly dancer undulating her body on one of the marble pillars that supported the tiled archways. The facilities would have graced any London gentleman's club. For the miners, the cost of using it would have been modest, being subsidised by the mines, but by the 1980s the mines were nearly all gone, the municipal grants had stopped and the running costs had to be met directly by the clientele, so it wasn't cheap to use. When I first went there, out of absolute curiosity, I was surprised to see how well it was still being used, but when I thought about it, there some pits were still running and, of course, there were the demolition crews for the steel works who continued to work there for several years in amongst the red dust and the blue asbestos. If you were a miner in work, you still had a good disposable income and a need to wash the grime of the day away. The baths also became a place to exercise the grey matter. The steam rooms became debating chambers and hotbeds (literally) of Arthur Scargill-style discussions about working conditions, employment, and the area's future.

There were three main steam rooms. The first was tepid. No one ever used it, except for lily-livered wimps. The second was a hot room for all those men who had weak hearts or the death rattle breathing that is a symptom of pneumoconiosis. The third room was burn-your-bollocks-off hot. This was the room for real men. When you had been in there for a while and had almost melted to a puddle on the floor, you took a trip to the plunge pool, located just outside the hot room and to one side of the massage cubicles. It was sited in a corridor and filled the space between the walls on each side, so there were no exit points. The walls of the pool sloped back under, making the floor of the pool bigger at the bottom in all directions, and it had smooth, rounded edges. If you tried to lower yourself in, you just fell backwards and then you were faced with a swim to the far end of the pool, where there were steps. There was only one way to get in, and that was to run and jump as far as you could. So why, you may ask, all the fuss about getting in? Because the water was just a little over freezing point and often had ice floating on the top. It was enough to stop your heart; in fact, on many of the occasions I jumped in, I think it did!

There was a shower in the massage room, and as a newcomer to the baths you would think to yourself that it would be a more comfortable experience. Oh! How wrong can you be? The shower looked like a free-standing, man-sized, *Sylvester and Tweety Pie*-style birdcage. You entered by a door constructed of inch-thick galvanised water pipes, and stood on two foot-shaped plates set in a floor criss-crossed with pipes and nozzles. The steel uprights of the cage were about three inches apart, and were studded with equally spaced nozzles. Above was the biggest shower rose you will have ever seen. It was about two and a half feet across; the modern trendy large roses pale into insignificance compared to it. Hanging down to about six feet off the ground was a brass loo chain with an ornate blue ceramic handle.

All the men in the place (as opposed to wussy ones) strutted their stuff, stark naked with the obligatory white towel slung over one shoulder, so, mimicking the rest I strode over to the shower, hung my towel over the hook provided and stepped naked into the cage. I swung the heavy gate shut behind me and then looked around for a tap or valve to turn. There wasn't one. So, what to do? As there was only the loo chain, I pulled it. Three things happened in quick succession. Firstly there was a loud "clunk" as the cage door was locked securely. Secondly, a motor started whirring somewhere in the bowels of the building and the pipe work in this iron maiden started to vibrate. Thirdly, freezing water erupted out of the omnipotent nozzles at the speed of Concorde. Every square inch of my skin and every orifice in my body seemed to be hit at the same time. The only place that escaped was the soles of my feet. At first, I could not tell whether it was hot or cold, but I must say that uncertainty did not last long. It was so shocking I could not scream, and if I had been able to, I think that if I'd had my mouth open, I would have drowned. At the end of what seemed like several hours of torture (in fact it was only about two minutes), everything stopped and the gate swung open with a loud "clang."

That was when a fourth thing happened – the assembled audience cheered.

It was quite a thing to witness. Even though I thought I was pink from the Turkish hot room, the water jets had all but removed my epidermis, making me look like a peeled Victoria plum.

After visiting the place for a few weeks, I thought it would be a good idea to try a massage. At least, that was what it was called on the plastic lettered pegboard in the reception area. It wasn't expensive so I thought why not? I had forgotten the fact that this was an establishment designed for getting real men clean and not some exclusive health spa. The wood partitioned massage cubicles were like stables, but with thick marble tables in them instead of horses. The fantasy that my masseuse might be some attractive nubile young female quickly evaporated when an Oddjob look-alike arrived. He carried with him a bottle of Fairy Liquid and the biggest loofah in the world. After the heat of the baths I was looking forward to the cold marble. Again, I had got it wrong. These tables were heated, and not just a little bit; they felt hot enough to fry an egg. The masseur kept his victims – sorry, clients – cool with what can only be described as a fire hose of cold water. That was where comfort ended. He squirted me all over with Fairy Liquid, and then gave me a good scrubbing from head to toe with his loofah. It took a maximum of ten minutes. It was at this point I was beginning to think perhaps this place was more suited to men with S and M proclivities.

Even though I often wondered why, I pretty much went at least one evening a week, perhaps for macho escapism or just washing the week's pressures away. I certainly was the cleanest I had been in my life, but it would have been nice to have been left with a layer of skin at the end.

Cut-throat partners, blaggards and thieves

November 6 1985 is a date imprinted on my memory. It was very early for snow, and, after I had driven over Stanhope Moor to the dairy through a thick white carpet of it, I was surprised to receive a telephone call from a Mr Stavely, who told me he was the northern area controller for Dairy Crest, a dairy company owned by the Milk Marketing Board. Mr Stavely invited me to their Durham Dairy to discuss what he described as "an interesting proposition." I went, full of curiosity, and was shown into a windowless, oak panelled boardroom. At the end of the long, highly polished table was an overweight businessman dressed in a stained brown suit and a tie decorated with scraps of a previous meal. This was Mr Stavely. He had a large, overfilled manila file in front of him and an expression somewhere between a Cheshire cat and The Joker from the *Batman* comics.

'Come in Phil,' he said from his leather stuffed executive chair. 'Make yourself comfortable.' All the other chairs were hard backed, highly polished and highly uncomfortable-looking wooden ones. I installed myself in one beside him.

'First of all, Mr Stavely,' I said. 'My first name is Philip, and secondly you can call me Mr Dixon until I decide whether or not I want to be on first name terms with you.' I was playing hardball, but he must have thought me a joke.

'I'll be quite candid,' he said. 'I am in need of a dairy manager for my Durham plant and I want to offer you the job.'

That took the wind out my sails. 'I have a job,' I said. 'And why on earth would you choose me?'

'I like what I've read, and I need a good man.' He said.

I was taken aback by this. 'You like what you read? You don't know me from Adam.'

At this he gathered up the file and slapped it down in front of him. He kept his hand firmly pressed down on the top preventing me from looking at it. 'I don't think you appreciate how much I do know about you.' He said.

I didn't think twice. I attempted to pull the file towards me, but he pressed down hard on it. The weight of the big man's arm was colossal, but he obviously hadn't read the bit about me walking bulls. I pulled his arm back. He yelped and let go. I started thumbing through the file while he nursed his arm (bullies are such wimps) and I was truly alarmed by the amount of information he had compiled about me and the financial detail of Merry Dale.

'Where on earth did you get this information?' I said.

'We like to know who our competitors are,' he said. 'It's taken a bit of time to find this out, and if you know what's good for you, you'll accept my offer.'

'Do you really think threatening people is the way to recruit them?' I said. 'I have a job and a business. I am not in the market for another.' And then I walked out. I was completely puzzled, and to be honest, I was flattered anybody would go to so much trouble to find out about me. Was it my conceit that stopped me from looking for another motive, or was the scale of the treachery beyond my ken?

Despite all the trade embargoes that had been placed in our way, Merry Dale had grown so well that the next phase of development, to bring dairy farmers into the business as partners, was brought forward by a year. I had just arranged my first meeting with the local dairy farmers. With them as partners, control over decision making would be lost to the original board, i. e. Noel, myself, Dave Dod and Bob Elliot, but the farmers would bring capital and business growth. Financial independence from local authority backing and banks would be just months away.

Then D-Day arrived.

Our wagon was making his deliveries, the truck collecting the bottled milk for Dod and Elliot milk delivery services duly arrived – and then, all our supplies stopped; no cartons, no bottles, no milk – nothing. By midday I was completely puzzled. By 4pm I was angry. Nobody was answering my calls. I was at a loss what to do, so I started ringing my customers to warn them that the next day's delivery was looking difficult. I was told it did not matter, as they had already had their delivery for the next day. It was then that the penny dropped and I realised we had been shafted, good and proper. A liquidator walked into the building at 6pm and insisted I handed over the keys, and that was it. Merry Dale was no more. I was shell-shocked and didn't know what to do. Just as I was leaving the building the liquidator called me back in for a telephone call (looking back on it, I don't remember hearing it ring, which is strange because we had installed a loud bell in the processing room so that we could hear, above the noise of the plant if we got a call). It was the Co-operative Dairy at Blaydon and the chap on the end of the phone said, 'I've heard on the grapevine of your troubles and wonder if you would be interested in being one of our production managers?'

I was speechless.

"If you're interested, give us a call,' he said. 'I'm very sorry for what's happened.' And then he hung up.

It took Noel and I many months to figure out exactly the extent of the betrayals, and of course we were penniless, so we couldn't pursue anybody in court, nor did we have any concrete evidence to prove collusion. I had no choice but to take the job at Blaydon. I had a mortgage to pay and a wife and three children to support.

Although I don't have evidence that would allow me to point a finger at any individual, group of individuals, or company, with hindsight I believe I had been the victim of the work of a cartel of the dairy companies in the area. As in all these things there are the outright villains and those in cahoots with them. Who could have supplied all the information Stavely had in his manila file? Only Noel, myself and Messrs Dod and Elliot had it to hand and it wasn't Noel who was the mole. Money is nearly always at the bottom of business treachery, so who could possibly have gained from Merry Dale's removal from the market place? With us gone, the north-eastern dairy industry lost a real threat to their status quo, and individual dairies gained the trade we lost.

Normally when successful businesses are bought and sold the goodwill is traded and has a value. Noel and I did not get a penny for this trade when another dairy took over supplying Messrs Dod and Elliot. The pair even started delivering to the rest of the customers who we had found long before they had joined the company. The total market value of all our trade must have been in excess of £200,000. Then there were the liquidators; at the point of liquidation there were sufficient funds to cover a payout of 95p in the pound (that's excluding the value of the goodwill) but at close of play they paid less than 50p in the pound. Next was the dairy industry, which had the power to control the price of milk, and whose members were able, should they choose to do so, spend a fortune to prevent a competitor getting started.

Last but not least was Dick Dastardly, who was paid three times: once by the people of Consett to look at my project, once by the cartel for information about Merry Dale, and once, afterwards by a large dairy farm that tried to mimic our idea.

Noel and I had received a hard lesson about the wider world.

With a family to support, I swallowed my pride and took the job of production manager at Associated Cooperative Creameries in Blaydon, which had been offered to me on that final day of Merry Dale. Because money was tight, the Saab had to go, and I bought a green Renault 12 estate that felt about 300 years old, and had a rear suspension made of jelly.

One advantage of working for someone else is that I had the luxury of paid holidays. We decided one year to go to the Gower Peninsular in South Wales. As we left with all our kit in the back, the rear end of the car was so low we could not get out the castle gates without it trailing on the ground. We stopped and redistributed the weight, putting all the heavy stuff on the back seat, as near to the front as possible, and the children, being the lightest, at the back in the boot area. It was a very slow journey but we eventually arrived at our campsite, and installed ourselves in a very "Ging Gang Goolie" sort of way with three dormitory tents and a larger mess tent. Perhaps I had overdone the kit a little, but at least we were comfortable in the rain. And for the first three days, boy, did it rain. We woke up on the fourth day to clear skies so we decided to walk to Three Cliffs Bay to spend the day playing in the waves.

No sooner had we arrived than a commotion started amongst the families who had settled there before us, who had gathered by the water's edge. Being naturally nosy, I went down to find out what was going on. A family – a man, a woman and two children – had been washed out to sea by a freak wave. The woman had managed to swim to shore, but her husband and the two children were still missing and, as you would expect, the woman was inconsolable.

As I stood on the sidelines and watched the drama unfold, somebody found the children playing in the shallows about 200 yards down the beach. Mother and children were tearfully united, but after a few moments she remembered her husband.

The alarm went up again to find him and we all ran around like headless chickens, with everybody being very British and not wanting to push themselves forward to take charge and organise a proper search. As a consequence he wasn't found until a group of boy scouts hiking over the cliff tops spotted him, floating about 200 yards off shore. Two of us swam out to pull him in and as soon as we got into the shallows I turned him to start mouth to mouth as I had been taught on my farm first aid course at Kirkley Hall. I was horrified with myself when I found I couldn't do it; not because of any squeamishness but since I had smashed my arm, I hadn't worked out how to do it one handed. I stood up, took a step backwards and said someone else would have to take over. I felt such a dick! After ten minutes the rescue helicopter arrived and whisked him away and we were told later he was alive when he was taken on board, but tragically he died before they could get him to hospital.

For the rest of the holiday, in the quiet times, I mulled over my ineptitude and I resolved if ever I came across a situation like that again I would take charge and organise. Why? Because I can, and in circumstances like these, people's opinion of you are unimportant. The opportunity to put my resolve into practice arrived sooner than I could possibly have expected.

I was driving into work the following week when a horrendous accident happened in front of me. The driver of a 20-tonne artic wagon lost control on a sharp corner and the lorry slewed across the road, crushing a transit van and its driver against the crash barrier. In the busy rush hour traffic, queues of impatient motorists quickly developed and a few people got out of their cars to help, but stood back, uncertain how to proceed. I got out and looked at the situation, and with the man at Three Cliffs Bay still preying heavily on my mind I took charge and started gelling people into action. To my amazement, that was all it took. People were eager to help and had good suggestions but were too timid to just go ahead and do them. A very young off-duty nurse volunteered to attend the van driver and with a reminder about the recovery position she rose to the challenge. By the time the police and the ambulance arrived we had the local GP, who lived three houses away, tending the van driver and the lorry driver, who were sitting beside the truck, drinking hot, sweet tea. The traffic was flowing, all be it slowly, under the direction of my newly appointed traffic controllers.

That evening, when I was passing through Stanhope, I decided I could do with a Turkish bath, but when I arrived at the hot room, sitting there like some fetid Buddha, was Angelo Marazzo, our *Mafioso* friend, surrounded by a bunch of young six-packed thugs. They were talking business amongst themselves. This was just too much. The man was invading my space. It was time for me to leave, so I went back home and the next day put The Beeches on the market.

The estate agent thought, as I did, that we would struggle to sell with our infamous neighbour hanging around like a spectre at the feast. Ideally, what I needed was a buyer who was more famous than Angelo, who just saw him as a colourful character. Sometimes in life, but not very often, a wish becomes a reality. Mine came in the form of children's television presenter Muriel Young and her husband Cyril Coke. Muriel was born in Bishop Middleham, not far from Weardale, and was returning to her roots. They had recently sold their five bedroom thatched house in Hampshire and they bought my apartment with their loose change. Cyril was a television producer and at that time was working with television journalist Ludovic Kennedy on his *Rough Justice* series. Sadly, Muriel nicknamed Ludovic "Ludo" and Cyril, "Cocoa." She had just retired from a lifetime of acting, producing and being the face that launched children's television on Associated-Rediffusion around 1955. At the age of 63 she was still a very talented, charismatic and attractive woman, but (there is always a but) she could not cook.

We were invited to a few of their dinner parties, the conversation and repartee being fascinating to a parochial Geordie like me. Stories of life in the media, "breaking new boundaries," and righting wrongs were all digested along with tinned Fray Bentos meat pies and burnt peas. How she managed to burn peas is a mystery to me but I'm afraid she did. Muriel and Cyril did find Angelo colourful but I always wondered how the pasta-loving Italian got on with the burnt peas.

On selling The Beeches we found a house that suited our needs. It was an end of terrace Victorian property in Saltwell View, which was in the Jewish quarter of Gateshead. The garden was tiny, but it overlooked the prize-winning Saltwell Park; 55 acres of open ground, studded with majestic specimen trees. It was hardly the countryside, but it would have to do for the time being, and it was a novelty having all amenities on hand. The big drop in commuter mileage was welcome too, as Blaydon, where I was working, was a five-minute drive away.

Life seemed to be settling back down when Carol started custody proceedings. It had been four years since she had left the children and me and although it took me a while to accept she had fallen out of love with me, what had always puzzled me was how she could leave her children. In the first year of our separation she only saw them a handful of times, and these visits were always sponsored by me, so it came as a complete surprise when the notification arrived for a custody suit. If our roles had been reversed and it had been me who had abandoned his children for that length of time, then I am sure in the 1980s it would never have got to court (in fact I'm not sure in these enlightened days it would get very far) but it seemed that she was acting on the basis that as I was male I was obviously incapable of caring for and cherishing my children. A three-day court hearing was set.

In the meantime, my working life continued at Associated Cooperative Creameries. This was another world for me. The factory, with its acres of concrete, was totally industrial and so different from the farms where I had spent the previous two decades of my life. The atmosphere was filled with steam, which belched out of grates as bottle washers jettisoned the scalding hot surplus caustic detergent. The noise of cleaned bottles, rattling their way at a rate of 600 a minute to the fillers was deafening, but the sound of the foil capping machine that stamped out the closures was beyond bearable. And that noise came from just one of the bottling lines. There were two more of them, plus the processing and cream potting plant. Ear defenders were essential. Not only did it seem very alien, but if I am honest I was totally out of my depth. Fortunately I was paired up with one of the other production managers. Her name was Karen and she had been told where I came from. She was very patient and kind to me as she taught me my role but it still took a good six months before I felt vaguely confident.

The production team was made up of four production managers, of which I was one, under a senior production manager. Our job was to oversee the day-to-day running of the dairy. The senior production manager, Grant McKechnie, rarely showed his face unless things went seriously wrong. Just before Christmas one year, the system went wrong big time and Grant spent all morning talking with head office managers. As soon as they left, Grant stormed out of his office and bustled his way to the one and only entrance to the site. There, he stamped his foot hard down on to the giant steel weighbridge, his ginger hair and beard trembling with the boiling rage he was keeping suppressed. He was saving its full force until he was face to face with the security guard who manned the almost-fortified entrance.

The representative of the security company came out of his cabin, putting on his hat to complete his uniform, to face six foot six of Scottish manhood. The guard's body had been abused by too many pints of Newcastle brown ale and his stained police style shirt was gaping open, its buttons having given up the fight against the blubber that was trying to escape its confines. Standing in front of Grant, who was dressed in the pristine whites of a senior dairy production manager, he looked like a slob.

'See yuu!' Said Grant. 'Whare's ma grundies?'

The security man was having none of this behaviour. 'Div'n't ye cum 'ere, shootin' yer gob off, ye heathen git!' He said.

For an outsider to comprehend an argument between a Glaswegian and a Geordie in full flight is all but impossible. Nor could either of them understand the other, but both their danders were up and the scale of the contretemps increased exponentially, with uncomprehending frustration driving it on to the inevitable conclusion. Grant clenched his fists and swung back for a blow, but a Geordie kiss to the nose took the ginger giant by surprise and he landed flat on his backside. The surrounding onlookers cheered, but had the common sense to separate the duelling trolls.

The cause of the matter was that, apparently, 3,800 litres of whipping cream had disappeared over a two-month period and the production team, of which McKechnie was head, was being held accountable. They pumped the cream into 400-litre stainless steel tanks on wheels, known as grundies, but sixteen had managed to leave the site, apparently whisked away by magic. It was meant to be impossible, because of the all-new, singing-and-dancing security system that was operated jointly by the dairy and the independent security firm. The security people checked and weighed each vehicle as it left the site's only exit. What made matters worse was that when the grundies were counted, there appeared to be none missing. Whoever was doing the thieving was returning these stainless steel Daleks a few days later, washed and ready to be filled again.

Each time the senior managers made a fuss, the missing grundy mystery would stop, but whenever the production of cream increased, the approach of a bank holiday, Christmas or Easter being the busiest the Bermuda Triangle of whipping cream would start up all over again.

The mystery was never solved, but new hygiene regulations banned the use of the reusable ex-brewery stainless steel containers in favour of the one trip 1000-litre Pallecon system. A Pallecon consisted of a throwaway plastic liner, protected by a collapsible shell, mounted on a wooden pallet. They literally weighed a tonne and were difficult to move without a forklift truck. Clearly the manoeuvrability of the grundies on their four large wheels was the reason for their theft, and seems to point to the fact that the thieves moved them by hand. But how they achieved this, when everything, including personnel, had to pass across the weighbridge, remained a mystery.

The other aspect to this conundrum was, who on earth bought all the cream? You couldn't sell 400 litres of cream on a market stall, and of course it only had a few days' shelf life. It must have been sold to a bakery that made an awful lot of cream cakes, or a factory making multi-layer desserts. We never found out who was taking it or who was buying it. When you employ 250 people, somebody normally comes up with a theory as to why it happened, but there was not even a rumour.

As well as theft on a large scale, petty theft at the dairy

was prevalent and we regularly caught individuals trying to smuggle out milk or pots of yogurt. They were always dismissed on the spot but it still didn't stop people trying, and all for a couple of pounds' worth of goods.

"Tennis Ball" was the exception.

He was a skinhead thug in his mid-thirties, and being squat in stature and with no neck, his nickname suited him down to the ground. Immediately after work he would go the boxing club on Scotswood Road and train for two or three hours, and then go down to the pub for several "Newky Broons" (Newcastle Brown ales). When he was tired, or had spent the housekeeping money, he would stagger his way home to beat his wife, 'Cos ye na, me supper wus clay cad' (because, you know, my supper was clay-cold) or any other excuse. If it wasn't the wife, it was the kids that got a beating. He was a nasty, nasty excuse of a man. Every member of staff was terrified of him, and although we suspected he was bullying people, the fear of retribution made stool pigeons thin on the ground. Behind the managers' backs he was full of it, but if a manager caught him out he would turn into Uriah Heep, being "ever so 'umble," but his unblinking crocodile eyes told a different tale.

At the rear of the production halls were a series of skips, containing the detritus from the production lines. Behind them was a steep grassy slope that led downhill to a three-metre high chain link fence topped with two strands of barbed wire. The slope continued down after the fence to the edge of the A695, which was a fast moving dual carriageway, designated a clearway.

Over a period of weeks I noticed trays of yogurts and cream, all of which seemed to be quite saleable, appearing on a Thursday or Friday in one corner of the skip nearest the fence. When I started monitoring this, I noticed different people placing this booty carefully on the skip, and even protecting it with a sheet of cardboard. It didn't always disappear straight away. Sometimes it stayed there until the skip was emptied the following week, but sometimes between the last five minutes of production and five minutes after the production staff finished, it would disappear.

I assumed somebody was going out the back of the dairy while the world and his wife were rushing to the car park to go home, and throwing the trays over the fence to a waiting accomplice. My plan to find out what was happening and put a stop to it was simple. I would park up on the side of the A695 with my hazard warning lights on and my bonnet up, pretending to have broken down, and watch what was going on. What happened next was the sort of thing that the makers of a Hollywood comedy would have been proud of.

Fifteen minutes before the end of the day I collected my old green Renault estate and drove out onto the dual carriageway. As I approached my planned observation position I put my hazard warning lights on and started to slow, intending to stop just beyond where I thought the yogurt exchange would occur.

To my surprise, who should appear at the top of the slope but Tennis Ball, with a stack of trays in his arms, his chin holding them in place. He ran out, stepped onto the top of the barbed wire, which was set down the slope, he then jumped down, and hit the remainder of the bank at a run. Looking at it afterwards, it wasn't that difficult. All it took was confidence and agility. The fence had been designed to keep people out, not keep them in, but having said that, I would not have dared do it. I reached across and opened the passenger door, intending to get out, as the driver's side traffic was too heavy. As I did so, Tennis plonked himself down in the seat with his contraband still firmly in his grasp, slammed the door and said, 'go!'

I went, and as we approached Scotswood Bridge the penny dropped with him and he turned to me with a look of horror and said, 'Where ye gannin?'

'Back to the Dairy! I said. 'I think I have to say you're nicked!'

Then, with me driving at thirty miles an hour, he just got out of the car and rolled onto the pavement with his booty. By the time I stopped he was up on his feet, throwing all the trays into the River Tyne. Then he legged it in the opposite direction.

On returning to the dairy we phoned the police, but to our dismay they said that it couldn't be theft if he took property with a negative value, and clearly if it was in the skip it was waste. We had no proof that he had arranged for the merchandise to be placed in the skip. The next day when he reported for work he was swaggering. The only discipline he received was a written warning for leaving the site by an unauthorised point. We couldn't even prove someone else had clocked out for him.

That day he was working on the bottling line, helping to feed the de-stacking machine with towers of dirty crates and bottles. His machine kept stopping and, as was his belligerent way he claimed that a chain was sticking. A fitter from the maintenance department was called out to fix it and Tennis Ball used this downtime to mouth off to anyone who would listen about the previous day's mischief and how the management couldn't touch him.

At the end of the day, when the line supervisor told him to stop putting any more crates through, he decided to kick a stack over. Unfortunately for him, I was standing behind him when he did this. I told him to sweep up the broken glass before he went home. He said, 'yu can stick the brush sideways up y'r arse.'

Disobeying a manager's direct order is gross misconduct and dismissal is immediate. I repeated myself. 'Sweep the glass up and tidy up the crates.'

'F*** off, ye posh git!' He replied and swaggered off, because he thought he was untouchable.

How wrong could he be? He was allowed on site the next day to be summarily dismissed, then frog-marched out of the front gate. As he was ejected, he seemed to physically shrink as the consequences began to sink in. Unemployment in the North East at that time was running at 25 per cent and he wasn't about to get a reference from anybody at ACC. Unfortunately for them, it was probably his wife and children that would suffer the most.

Later that week when I had finished my shift at the treadmill that was ACC and was making my way back to Saltwell View, imagine my surprise when I saw a very familiar looking blue Rolls Royce parked on the double yellow lines immediately outside our house. There, on the doorstep when I rounded the hedge, I saw none other than Angelo. My heart sank. What on earth does he want? I thought. But I was wrong to expect trouble. He apparently thought of himself as a mate. When that leap had occurred I'm not sure; maybe when I helped him into his house after he had been stabbed, but Angelo had heard of my troubles with Carol and he told me he thought, as I did, that if a man had abandoned his children for four years, there would not have been a battle. The application for custody would have been thrown out of court and it bloody well wouldn't be a three-day hearing. He was Mr Angry again. It must have come from the Italians' love of family and children. He was incensed at the sex discrimination in the British justice system and he asked me if I wanted him to sort out the ex and her man.

'What do you mean by "sort out?"' I asked.

'Hospital,' he said, 'or more, if you want.'

'Er, very tempting,' I said, 'but no thank you. I am sure it will resolve itself with a bit of hard work.'

'If you're sure,' he said, 'you've only to ask.'

With that bombshell I thanked him again, bid him good evening and closed the front door behind me.

'BLOODY HELL!' I thought as I leant on the door. 'Did that really happen?'

He had timed his visit well. We had just been notified that the hearing was to be held in Huddersfield, near to where Carol lived, which was 120 miles away from Gateshead. Society's leeches had already been engaged; lawyers on both sides would be digging the dirt, desperate to discredit one another. Child psychologists, social workers, experts in this, witnesses in that, all drawing their fat fees, all paid for by legal aid. It was just plain crazy.

The day dawned all too quickly. We waited in the antechamber, each side facing one another but not daring to commit to eye contact. Both Carol and I seemed to have aged noticeably. The clerk of the court called our names and into the traditional courtroom we all filed, taking our prescribed seats to wait for the judge, who was not wigged and gowned but still managed to have the air of a lord high executioner. As he swept in via a little side door next to his bench (which was more like a throne on high), we all leapt to our feet. We were told to sit, and battle commenced.

The judge just sat on his bench and listened, never giving anything away and rarely asking any questions as the myriad of witnesses paraded in front of him. Just as it was winding up on the second day, the opposing lawyer stated that the children were to be within the court precincts the following day, because it was a distinct possibility that Sharron, as the eldest, could be asked with which parent she would like to live. I was horrified but was told by the judge I had to comply.

I agonised all night about what I could possibly do to prevent this travesty. It had already been decided, quite rightly, that the children should not be split up. How could the court make an eleven-year-old responsible for deciding where she and her two siblings were to live? Which parent she chose was not the issue. We were the adults and we had to take the responsibility for the decision. The only solution I could think of was if Sharron was asked to take the stand, then I would have to concede custody, thus taking that decision away from her.

When the morning arrived and I sat in that austere place, I felt as if my insides had been sucked out and I had shrunk to half my former size, but I had a hard-knotted resolve sitting in the base of my stomach, lying there like a block of Semtex waiting for a wrong decision to explode it. Yet again, it looked as if I would have my world destroyed. The wheels of the legal system continued to grind on, the minutes seeming like hours, and then, as if someone had pushed a fast-forward button, the lawyers were finished and all attention was focused on the judge. He did not retire to consider his decision. He pitched straight in, condemning the suggestion that an eleven-year-old should be asked to take such an important decision. He then went on to deliver effusive praise on me and the stable home I had set up under such difficult circumstances. In his next breath, he granted me full custody of the children. He finished up with the declaration that if there was ever to be any more court proceedings involving my family it was to be heard in front of him.

The relief flooded into my re-inflating body, the knotted cord untying and a secure serenity returning to the only truly important part of my life. As I drove back up to the North East with my children cocooning me I resolved I must find a way to take them all back to rural farm life where I would be around them all day, every day. Then an old thought occurred again, the same one that crossed my mind after I had lost the use of my arm, that if a load of nappy-clad Egyptians could build the pyramids, there had to be a way to get back to my farming dream.

But that's the next story to be written.

About the author

Philip Dixon's formative years were spent around the banks of the River Tyne, in the north of England and later in the 1950s and 1960s on farms in the area. He describes himself as "first and foremost a dairy farmer" and has spent much of his working life farming and managing dairies in England and France. He has also owned and run a chambre d'hote in France.

Besides farming, Philip has renovated houses, appeared on television and radio, and had articles written about him in *Farmers Weekly*. Presently he lives in Surrey with his wife, Heather. He has four children of his own and is stepfather to three more, all of whom are now adults. He has recently returned from Kenya where he was working on a dairy project.

Follow Philip Dixon on Facebook.
(facebook.com/farmerphilipauthor)